Synthesis Lectures on Mechanical Engineering

This series publishes short books in mechanical engineering (ME), the engineering branch that combines engineering, physics and mathematics principles with materials science to design, analyze, manufacture, and maintain mechanical systems. It involves the production and usage of heat and mechanical power for the design, production and operation of machines and tools. This series publishes within all areas of ME and follows the ASME technical division categories.

Kamran Hassani · Farhad Azimifar

Introduction to Continuum Mechanics

Springer

Kamran Hassani
School of Mechanical,
Industrial, and Aeronautical Engineering
University of Witwatersrand
Johannesburg, South Africa

Farhad Azimifar
Department of Biomedical Engineering,
Institute of Artificial Intelligence and Social
and Advanced Technologies
Isf.C., Islamic Azad University
Isfahan, Iran

ISSN 2573-3168 ISSN 2573-3176 (electronic)
Synthesis Lectures on Mechanical Engineering
ISBN 978-3-032-05075-5 ISBN 978-3-032-05076-2 (eBook)
https://doi.org/10.1007/978-3-032-05076-2

This Springer imprint is published by the registered company Springer Nature Switzerland AG
The registered company address is: Gewerbestrasse 11, 6330 Cham, Switzerland

Introduction

The existing physical theories note that the materials are discontinuous on a microscopic scale and a material made of atoms, molecules, and other components. In continuum mechanics, we intend to investigate material problems and focus on its performance, noting to material deformation and movement. This means that we do not study each of the components that form the material, and we aim to review the resultant deformation of the material. This is called macroscopic analysis of material. In continuum mechanics, the mechanical behavior of fluids and solids is studied on a macroscopic scale and assumes that the material is completely continuous. It is assumed that a continuum environment has a uniform distribution of material, and quantities like density, displacement, velocity, etc., are defined based on spatial continuous functions in different three-dimensional spaces. Therefore, we assume that an infinitely small volume of material is continuous, and the atoms' behavior is not considered separately. Thus, the material is a continuous body without atoms. Apart from the assumption of continuity, we consider two other important assumptions in continuum environments: the identical properties in all locations, which are referred to as homogeneity, and isotropy, which refers to directionally invariant vector properties in a body.

There are two categories of variables in continuum mechanics: one category is dependent on the force, and the other is dependent on movement parameters, including velocity, acceleration, and displacement. It should be noted that continuum mechanics deals with deformable bodies only. A rigid body is also considered a continuous environment because the rigid body is ideal, and all bodies experience deformity when they are under force exertion. In deformable bodies, the relative velocity of the particles relative to each other is important, and this relative velocity is the main cause of deformation, displacement, and velocity in a body.

There are two types of equations in continuum mechanics:

- The equations that apply to all materials, including mass conservation and energy equations. These equations are representative of universal physical laws.
- The constitutive equations that review the mechanical properties of specific materials.

To sum up, continuum mechanics is a branch of mechanical engineering science that can describe the concepts of stress and strain in solids, fluids, and gases under different loadings. It is assumed that the molecular structure of a material is continuous, and there is no free space between the atoms. Furthermore, there is a uniform material distribution in a specific space that is a continuous environment.

Contents

1 Mathematics

The first step for the expression of continuum mechanics concepts is to be familiar with the required mathematical methods. Thus, we start with the matrix concept, as many engineering students have enough preliminary knowledge about matrices. A $m \times n$ matrix is presented as below:

$$A = \begin{bmatrix} T_{11} \ T_{12} \ \dots T_{1n} \\ T_{21} \ T_{22} \ \dots T_{2n} \\ \vdots \quad \vdots \\ T_{m1} T_{m2} \dots T_{mm} \end{bmatrix} \tag{1.1}$$

m refers to the number of rows and n to columns. Normally, we deal with 3 × 1, 1 × 3, and 3 × 3 matrices. 3 × 3 matrix is called a square matrix and is shown as:

$$A = \begin{bmatrix} A_{11} \ A_{12} \ A_{13} \\ A_{21} \ A_{22} \ A_{23} \\ A_{31} \ A_{32} \ A_{33} \end{bmatrix} \tag{1.2}$$

The transpose of a square matrix is defined as below:

$$A^T = \begin{bmatrix} A_{11} \ A_{21} \ A_{31} \\ A_{12} \ A_{22} \ A_{32} \\ A_{13} \ A_{23} \ A_{33} \end{bmatrix} \tag{1.3}$$

If $A = A^T$ The matrix is called a symmetric matrix, and the matrix is called skew-symmetric if $A = -A^T$ $A = -A^T$.

K. Hassani and F. Azimifar, *Introduction to Continuum Mechanics*, Synthesis Lectures on Mechanical Engineering, https://doi.org/10.1007/978-3-032-05076-2_1

The unit matrix is one of the most applicable matrices in continuum mechanics, which is defined as:

$$I = \begin{bmatrix} 1\,0\,0 \\ 0\,1\,0 \\ 0\,0\,1 \end{bmatrix} \tag{1.4}$$

The determinant of a square matrix is calculated easily by the Sarus method:

$$A = \begin{bmatrix} a_{11} & a_{12} & a_{13} \\ a_{21} & a_{22} & a_{23} \\ a_{31} & a_{32} & a_{33} \end{bmatrix} \rightarrow |A| = \begin{vmatrix} a_{11} & a_{12} & a_{13} \\ a_{21} & a_{22} & a_{23} \\ a_{31} & a_{32} & a_{33} \end{vmatrix} \tag{1.5}$$

$$\begin{aligned} |A| = &\begin{vmatrix} a_{11} & a_{12} & a_{13} \\ a_{21} & a_{22} & a_{23} \\ a_{31} & a_{32} & a_{33} \end{vmatrix}\begin{vmatrix} a_{11} & a_{12} \\ a_{21} & a_{22} \\ a_{31} & a_{32} \end{vmatrix} \\ &\equiv (a_{11}a_{22}a_{33} + a_{12}a_{23}a_{31} + a_{13}a_{21}a_{32}) \\ &- (a_{12}a_{21}a_{33} + a_{11}a_{23}a_{32} + a_{13}a_{22}a_{31}) \end{aligned} \tag{1.6}$$

Before starting the new mathematical concepts, the concepts of vectors are reviewed. The vectors are physical quantities that have size and direction. A vector(V) is presented as:

$$\vec{V} = x_i + y_j + z_k \tag{1.7}$$

In the introduction to continuum mechanics, we only refer to Cartesian coordinates, but many other continuum mechanics problems are solved using cylindrical or spherical coordinate systems. The magnitude and size of a vector are calculated by the following relation:

$$|\overrightarrow{V}| = \sqrt{x^2 + y^2 + z^2} \tag{1.8}$$

The vectors' summation has commutative and associative properties as per below:

$$\vec{a} + \vec{b} = \vec{b} + \vec{a} \tag{1.9}$$

$$\left(\vec{a} + \vec{b}\right) + \vec{c} = \vec{a} + \left(\vec{b} + \vec{c}\right) \tag{1.10}$$

In this part, we introduce direction cosines in vector algebra, that is used frequently in continuum mechanics. The vector $\vec{V}$ is defined in Cartesian coordinates with three components of V_x, V_y, and V_z. The vector is also defined in terms of unit vectors of i, j, and k. In the future, we must use the numbers 1, 2, and 3 instead of i, j, and k to present the unit vectors. Thus, we have:

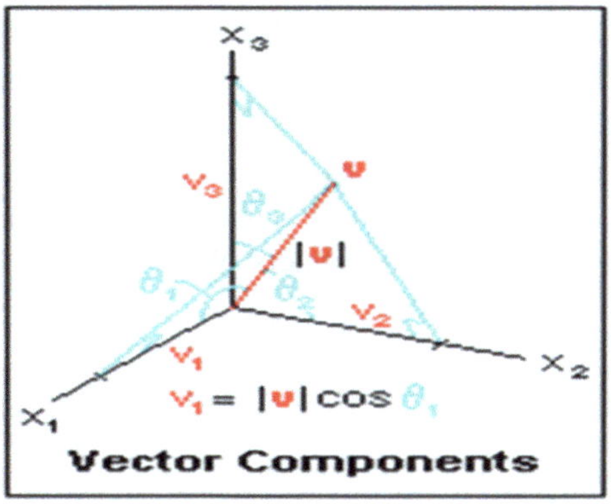

Fig. 1.1 The projection of a vector and related angles with the Cartesian axes

$$\vec{V_x} = V_{xi} = V_1.i_1, \vec{V_y} = V_{yj} = V_2 i_2, \vec{V_z} = V_{zk} = V_3 i_3 \tag{1.11}$$

Then we represent the vector $\vec{V}$ in a final form of:

$$\vec{V} = \vec{V_x} + \vec{V_y} + \vec{V_z} = V_{xi} + V_{yj} + V_{zk} = \mathrm{V_1 i_1 + V_2 i_2 + V_3 i_3} \tag{1.12}$$

Figure 1.1 shows the concept of direct cosines well, where $\mathrm{V_1 = V_x}$, $\mathrm{V_2 = V_y}$, and $\mathrm{V_3 = V_z}$ are the projections of $\vec{V}$ on the Cartesian coordinates:

$$\begin{aligned} V_x &= VCos\theta_1 = \mathrm{V_1} \\ V_y &= VCos\theta_2 = \mathrm{V_2} \\ V_z &= VCos\theta_3 = \mathrm{V_3} \end{aligned} \tag{1.13}$$

$\mathrm{Cos}\theta_1$, $\mathrm{Cos}\theta_2$, and $\mathrm{Cos}\theta_3$ are the direction cosines of $\vec{V}$ along with the Cartesian axes. Therefore, we have:

$$V^2 = Vx^2 + Vy^2 + Vz^2 = V1^2 + V2^2 + V3^2 \tag{1.14}$$

Now, the relation (1.13) is substituted in the above one, and we have:

$$\begin{aligned} V^2 &= (VCos\theta_1)^2 + (VCos\theta_2)^2 + (VCos\theta_3)^2 \\ &= V^2\left(Cos\theta_1^2 + Cos\theta_2^2 + Cos\theta_3^2\right) \end{aligned}$$

$$Cos\theta_1^2 + Cos\theta_2^2 + Cos\theta_3^2 = 1 \tag{1.15}$$

This relation (1.14), is the key formula, and it is used in the following chapters.

Indicial Notation and Summation Convention

In continuum mechanics problems, we are confronted with too many multiplications of matrices or vectors. These operations take a long time and need special notice. Therefore, indicial notation has been introduced to shorten the calculation time of multiplications. Consider the vector $\vec{V}$ again, where i_1, i_2, and i_3 are used instead of x, y, and z coordinates, and they are unit vectors. We rewrite the vector definition as below:

$$\begin{aligned}\vec{V} &= V_1x + V_2y + V_3Z = V_1i_1 + V_2i_2 + V_3i_3\\ &\equiv \sum_{i=1}^{3} V_i.i_i \equiv V_i.i_i\end{aligned} \tag{1.16}$$

Now, see the following phrase:

$$x_i = A_{ij}.y_i \tag{1.17}$$

i is called the free index, and j is the dummy index. In index notation, an index that appears twice within a term is called a dummy index, implying a summation over that index's range. An index that appears only once in a term is called a free index, and it can take any value in its allowed range. The relation (1.17) is rewritten based on this definition as:

$$\begin{cases} x_1 = (A_{11} + A_{12} + A_{13})y_1 \\ x_2 = (A_{21} + A_{22} + A_{23})y_2 \\ x_3 = (A_{31} + A_{32} + A_{33})y_3 \end{cases} \tag{1.18}$$

Note that the summation is always done for 1 to 3, referring to the Cartesian coordinates.

Kronecker Delta

The Kronecker delta is a function that is used a lot in continuum mechanics as a simplifying symbol:

$$\delta_{ij} = \begin{cases} 1 \; i = j \\ 0 \; i \neq j \end{cases} \tag{1.19}$$

The relation implies that $\delta_{11} = \delta_{22} = \delta_{33} = 1$ and $\delta_{12} = \delta_{13} = \delta_{23} = 0$. The Kronecker delta concept is the same as the unit matrix:

$$I = \begin{bmatrix} 1 & 0 & 0 \\ 0 & 1 & 0 \\ 0 & 0 & 1 \end{bmatrix} = \begin{bmatrix} \delta_{11} & \delta_{12} & \delta_{13} \\ \delta_{21} & \delta_{22} & \delta_{23} \\ \delta_{31} & \delta_{32} & \delta_{33} \end{bmatrix} \tag{1.20}$$

Now, note the following useful examples:

$$\delta_{\mathrm{ii}} \equiv \delta_{11} + \delta_{22} + \delta_{33} = 1 + 1 + 1 = 3$$

(B)
$$\begin{aligned} &\delta_{\mathrm{ij}}.\mathrm{V_i} = (\delta_{11} + \delta_{12} + \delta_{13})V_1 + (\delta_{21} + \delta_{22} + \delta_{23})V_2 \\ &+(\delta_{31} + \delta_{32} + \delta_{33})V_3 = \delta_{11}V_1 + \delta_{22}V_2 + \delta_{33}V_3 = V_1 + V_2 + V_3 = V_j \end{aligned}$$

As seen in example B, if the Kronecker delta acts on a vector, then the free index is changed from i to j. Thus, the following relation should be memorized as one of the most applicable relations in the next chapters:

$$\delta_{ij}.V_i = V_j \tag{1.21}$$

The following example refers to another important relation:

(C)
$$\begin{aligned} \delta_{\mathrm{mn}}.\delta_{\mathrm{mp}} &= \left(\delta_{1\mathrm{n}}.\delta_{1\mathrm{p}} + \delta_{2\mathrm{n}}.\delta_{2\mathrm{p}} + \delta_{3\mathrm{n}}.\delta_{3\mathrm{p}}\right) \\ &\equiv (\delta_{11} + \delta_{12} + \delta_{13})\delta_{11} + (\delta_{21} + \delta_{22} + \delta_{23})\delta_{21} + (\delta_{31} + \delta_{32} + \delta_{33})\delta_{31} \\ &+ (\delta_{11} + \delta_{12} + \delta_{13})\delta_{12} + (\delta_{21} + \delta_{22} + \delta_{23})\delta_{22} + (\delta_{31} + \delta_{32} + \delta_{33})\delta_{32} \\ &+ (\delta_{11} + \delta_{12} + \delta_{13})\delta_{13} + (\delta_{21} + \delta_{22} + \delta_{23})\delta_{23} + (\delta_{31} + \delta_{32} + \delta_{33})\delta_{33} \end{aligned}$$

As other Kronecker are zero except $\delta_{11} = \delta_{22} = \delta_{33} = 1$ so we have:

$$\delta_{mn}.\delta_{mp} = (\delta_{11} + \delta_{12} + \delta_{13}) + (\delta_{21} + \delta_{22} + \delta_{23}) + (\delta_{31} + \delta_{32} + \delta_{33}) = \delta\mathrm{np}$$

Or:

$$\delta_{mn}.\delta_{mp} = \delta_{np} \tag{1.22}$$

We will use relation (1.22) many times in our further problems.

The Permutation Symbol

Another applicable symbol that is used in continuum mechanics is the permutation symbol:

$$e_{ijk} = \begin{cases} 1 & ijk = 123 = 231 = 312 \\ -1 & ijk = 132 = 213 = 321 \\ 0 & \textit{if two index are identical at least} \end{cases}$$

$$\begin{cases} e_{123} = e_{231} = e_{312} = 1 \\ e_{132} = e_{213} = e_{321} = -1 \end{cases} \tag{1.23}$$

One simple method is to consider the clockwise direction as positive based on the following image Fig. 1.2.

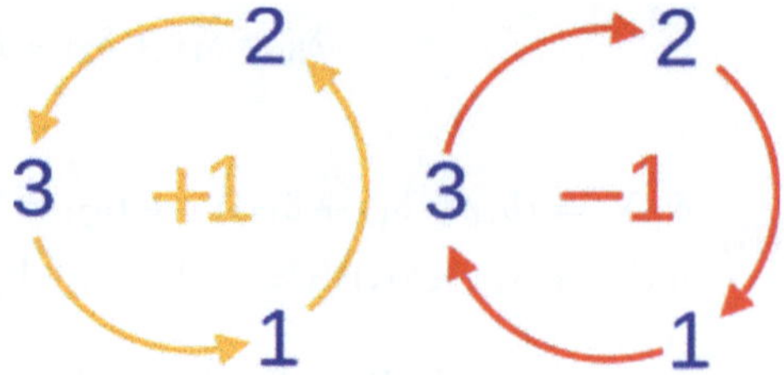

Fig. 1.2 The positive direction for determining the sign of the permutation symbol

The Relation Between the Kronecker Delta and the Permutation Symbol

The determinant of matrix A 3 ×3 is calculated by Sarus:

$$|A| = \begin{vmatrix} \delta_{i1} & \delta_{i2} & \delta_{i3} \\ \delta_{j1} & \delta_{j2} & \delta_{j3} \\ \delta_{k1} & \delta_{k2} & \delta_{k3} \end{vmatrix}$$
$$= \delta_{i1}\left(\delta_{j2}.\delta_{k3} - \delta_{j3}.\delta_{k2}\right) - \delta_{i2}\left(\delta_{j1}.\delta_{k3} - \delta_{j3}.\delta_{k1}\right) + \delta_{i3}\left(\delta_{j1}.\delta_{k2} - \delta_{j2}.\delta_{k1}\right)$$

If i = 1 then δ_{i2}, δ_{i3} are equal to δ_{12}, δ_{13} which are zero. The determinant of A is calculated as 1 for i = 1, j = 2, and k = 3. Moreover, the determinant of A is (−1) for i = 1, j = 3, and k = 2, and it is zero for other conditions. For example:

$$i = 1, j = 2, k = 3 \rightarrow$$
$$\begin{vmatrix} \delta_{11} & \delta_{12} & \delta_{13} \\ \delta_{21} & \delta_{22} & \delta_{23} \\ \delta_{31} & \delta_{32} & \delta_{33} \end{vmatrix} = \delta_{11}(\delta_{22}.\delta_{33} - \delta_{23}.\delta_{32}) \equiv \delta_{11}(\delta_{22}.\delta_{33}) = 1$$

Similarly, for i = 2 and i = 3, we see that the determinant of A is 0, (−1), and 1 for different values of j and k. On the other hand, we calculate once again the permutation symbol for the same values of i, j, and k as above:

i = 1, j = 2, and k = 3 then $e_{ijk} = 1$.

i = 1, j = 3, and k = 2 then $e_{ijk} = -1$.

i = 1, j = 2 and k = 2 then $e_{ijk} = 0$.

We conclude that the determinant of A for different values of i, j, and k is 0, 1, and (−1), which gives the same results for the permutation symbol. Thus, it can be deducted:

$$\begin{vmatrix} \delta_{i1} & \delta_{i2} & \delta_{i3} \\ \delta_{j1} & \delta_{j2} & \delta_{j3} \\ \delta_{k1} & \delta_{k2} & \delta_{k3} \end{vmatrix} \equiv e_{ijk} \tag{1.24}$$

There is another useful relation that could be derived from relation (1.24):

$$e_{ijk}.e_{pqr} = \begin{vmatrix} \delta_{i1} & \delta_{i2} & \delta_{i3} \\ \delta_{j1} & \delta_{j2} & \delta_{j3} \\ \delta_{k1} & \delta_{k2} & \delta_{k3} \end{vmatrix} \begin{vmatrix} \delta_{p1} & \delta_{q1} & \delta_{r1} \\ \delta_{p2} & \delta_{q2} & \delta_{r2} \\ \delta_{p3} & \delta_{q3} & \delta_{r3} \end{vmatrix} \tag{1.25}$$

Remembering from mathematics, we know that for two matrices A and B, whose determinants are $|A|$and $|B|$, then $|A|.|B| = |A.B|$. Moreover, the determinant of a matrix is equal to the determinant of its transpose, and we have used this property in relation (1.25) for e_{pqr}.

Relation (1.25) is rewritten as below:

$$e_{ijk}.e_{pqr} = \left| \begin{pmatrix} \delta_{i1} & \delta_{i2} & \delta_{i3} \\ \delta_{j1} & \delta_{j2} & \delta_{j3} \\ \delta_{k1} & \delta_{k2} & \delta_{k3} \end{pmatrix} \begin{pmatrix} \delta_{p1} & \delta_{q1} & \delta_{r1} \\ \delta_{p2} & \delta_{q2} & \delta_{r2} \\ \delta_{p3} & \delta_{q3} & \delta_{r3} \end{pmatrix} \right| \equiv \begin{vmatrix} \delta_{ip} & \delta_{iq} & \delta_{ir} \\ \delta_{jp} & \delta_{jq} & \delta_{jr} \\ \delta_{kp} & \delta_{kq} & \delta_{kr} \end{vmatrix} \tag{1.26}$$

We have calculated δ_{ip} and δ_{jp} for example, and other ones should be calculated by the readers as a practice:

The first row of the first matrix should be multiplied by the first column of the second matrix, which gives:

$$\delta_{i1}.\delta_{p1} + \delta_{i2}.\delta_{p2} + \delta_{i3}\delta_{p3} \tag{1.27}$$

Relation (1.27) is expressed as:

$$\delta_{iq}\delta_{pq} = \delta_{i1}.\delta_{p1} + \delta_{i2}\delta_{p2} + \delta_{i3}\delta_{p3} (\textit{the difination of dummy index}) \tag{1.28}$$

On the other hand, $\delta_{iq}\delta_{pq} = \delta_{iq}\delta_{qp=}\delta_{ip}\delta_{iq}\delta_{pq} = \delta_{iq}\delta_{qp=}\delta_{ip}$.

The Kronecker delta is like the unit matrix, which is a symmetric matrix; therefore $\delta_{pq} = \delta_{qp}$.

In the same method, we can calculate the result of multiplying the second row of the first matrix by the first column of the second matrix:

$$\delta_{j1}.\delta_{p1} + \delta_{j2}.\delta_{p2} + \delta_{j3}.\delta_{p3} \equiv \delta_{jk}.\delta_{pk} = \delta_{jk}.\delta_{kp} = \delta_{jp} \tag{1.29}$$

The following relations could be proved by readers as a practice:

(a) $e_{pqs}.e_{snr} = \delta_{pn}.\delta_{qr} - \delta_{pr}.\delta_{qn}$
(b) $e_{pqs}.e_{sqr} = -2\delta_{pr}$
(c) $e_{ijk}.e_{ijk} = 6$

The Vectors Products

1. The dot product:

Two vectors of $\vec{a}$ and $\vec{b}$ are considered with θ the angle between them. If $|\vec{a}|$ and $\left|\vec{b}\right|$ are defined as the magnitude (length) of the vectors, then the dot product is determined as:

$$\vec{a} = a_1 i_1 + a_2 i_2 + a_3 i_3$$

$$\vec{b} = b_1 i_1 + b_2 i_2 + b_3 i_3$$

$$\begin{aligned} \vec{a}.\vec{b} &\equiv |\vec{a}|.\left|\vec{b}\right|.\mathrm{Cos}\theta = (a_1 i_1 + a_2 i_2 + a_3 i_3).(b_1 i_1 + b_2 i_2 + b_3 i_3) \\ &\equiv a_1 b_1 + a_2 b_2 + a_3 b_3 \equiv a_i b_i \end{aligned} \tag{1.30}$$

Now, we calculate the dot products of unit vectors (i_1, i_2, i_3):

$$i_1.i_2 = |i_1|.|i_2|.cos90^{\circ} \equiv 1 \times 1 \times 0 = 0$$

$$i_1.i_3 = |i_1|.|i_3|.cos90^{\circ} \equiv 1 \times 1 \times 0 = 0$$

$$i_1.i_1 = |i_1|.|i_1|.cos0^{\circ} \equiv 1 \times 1 \times 1 = 1$$

Which results:

$$i_i.i_j = \begin{cases} 0 i \neq j \\ 1 i = j \end{cases} \tag{1.31}$$

Or:

$$i_i.i_j = \delta_{ij} \tag{1.32}$$

It should be noted that the unit vectors are orthogonal.

2. The Cross product of the vectors:

The dot Product gives a scalar (ordinary number) answer and is sometimes called the scalar product. But there is also the Cross Product, which gives a vector as an answer, and is sometimes called the vector product:

$$\vec{a} = a_1 i_1 + a_2 i_2 + a_3 i_3$$

$$\vec{b} = b_1 i_1 + b_2 i_2 + b_3 i_3$$

$$\vec{a} \times \vec{b} \equiv |\vec{a}|.\left|\vec{b}\right|.sin\theta = \begin{vmatrix} i_1 & i_2 & i_3 \\ a_1 & a_2 & a_3 \\ b_1 & b_2 & b_3 \end{vmatrix} \tag{1.33}$$

If we define the cross product based on indicial notation:

$$\begin{gathered} \vec{a} \times \vec{b} = \begin{vmatrix} i_1 & i_2 & i_3 \\ a_1 & a_2 & a_3 \\ b_1 & b_2 & b_3 \end{vmatrix} = \begin{vmatrix} i_1 & i_2 & i_3 \\ a_1 & a_2 & a_3 \\ b_1 & b_2 & b_3 \end{vmatrix}\begin{vmatrix} i_1 & i_2 \\ a_1 & a_1 \\ b_1 & b_2 \end{vmatrix} \\ (a_2 b_3 - a_3 b_2) i_1 + (a_3 b_1 - a_1 b_3) i_2 + (a_1 b_2 - a_2 b_1) i_3 \equiv e_{ijk}.a_j.b_k.i_i \end{gathered} \tag{1.34}$$

For the case that i = 1(free index) and j, k are dummy indexes:

$$\begin{aligned} i = 1 \rightarrow \; & e_{1jk}.a_j.b_k.i_1 \\ & = (e_{111} a_1 b_1 + e_{121} a_2 b_1 + e_{131} a_3 b_1 + e_{112} a_1 b_2 \\ & + e_{122} a_2 b_2 + e_{132} a_3 b_2 + e_{113} a_1 b_3 + e_{123} a_2 b_3 + e_{133} a_3 b_3) i_1 \\ & \equiv (0 + 0 + 0 + 0 + 0 - a_3 b_2 + 0 + a_2 b_3 + 0) i_1 = (a_2 b_3 - a_3 b_2) i_1 \end{aligned}$$

One of the mathematical relations that we face in the future is the volume of a parallelepiped that has been produced by three vectors of $\vec{a}$, $\vec{b}$, $\vec{c}$ and is calculated by:

$$\left(\vec{a} \times \vec{\mathrm{b}}\right).\vec{c} \equiv \begin{vmatrix} a_1 & a_2 & a_3 \\ b_1 & b_2 & b_3 \\ c_1 & c_2 & c_3 \end{vmatrix} \equiv e_{ijk}.a_i.b_j.c_k \tag{1.35}$$

Tensor Concept

A tensor is an algebraic object that describes a multilinear relationship between sets of algebraic objects related to a vector space. Tensors may map between different objects, such as vectors, scalars, and even other tensors. In the simplest form, a tensor has only one member, which is referred to as a scalar tensor. The vector is a tensor with two members, and the matrices are another type of tensor. Tensor is in fact a linear mapping that maps a vector like A to another vector like TA:

$$\begin{gathered} T(A + \mathrm{B}) = \mathrm{TA} + \mathrm{TB} \\ T(a.\mathrm{A}) = \mathrm{a.TA} \end{gathered} \tag{1.36}$$

Tensors form a vector space, and the summation, or the product is defined as:

$$(A + \mathrm{B})\mathrm{V} = \mathrm{A.V} + \mathrm{B.V}$$
$$(a.\mathrm{A})\mathrm{V} = \mathrm{a}(\mathrm{A.V}) \tag{1.37}$$

Which a is a scalar and A, V, B are vectors.

Dyadic Product
There is another form of vector product which results in a matrix and is called a Dyad:

$$T = a \otimes \mathrm{b} \rightarrow \mathrm{T_{ij}} = \mathrm{a_i} b_j \tag{1.38}$$

$$T = \vec{a} \otimes \vec{\mathrm{b}} = \begin{pmatrix} a_1 \\ a_2 \\ a_3 \end{pmatrix} \begin{pmatrix} b_1 & b_2 & b_3 \end{pmatrix} = \begin{pmatrix} a_1b_1 & a_1b_2 & a_1b_3 \\ a_2b_1 & a_2b_2 & a_2b_3 \\ a_3b_1 & a_3b_2 & a_3b_3 \end{pmatrix} \tag{1.39}$$

Dyadic products have not commutative property but associative:

$$a \otimes \mathrm{b} \neq \mathrm{b} \otimes a$$
$$a \otimes (\mathrm{b} + \mathrm{c}) = a \otimes \mathrm{b} + a \otimes c \tag{1.40}$$
$$(ma) \otimes \mathrm{b} = a \otimes (m\mathrm{b}) = m(a \otimes b)$$

m is a scalar.

Tensor Product
The tensor product is the same as matrices product:

$$C = \mathrm{A.B} = \begin{bmatrix} a_1 & a_2 & a_3 \\ b_1 & b_2 & b_3 \\ c_1 & c_2 & c_3 \end{bmatrix} \begin{bmatrix} d_1 & d_2 & d_3 \\ e_1 & e_2 & e_3 \\ f_1 & f_2 & f_3 \end{bmatrix} \tag{1.41}$$

For example, the first component of C matrix, first row and first column, is calculated by:

$$C = \begin{bmatrix} a_1d_1 + a_2e_1 + a_3f_1 & a_1d_2 + a_2e_2 + a_3f_2 & a_1d_3 + a_2e_3 + a_3f_3 \end{bmatrix} \tag{1.42}$$

The indicial notation of the matrix's product is shown as:

$$C = \mathrm{A.B} \rightarrow C_{ij} = A_{ik}.B_{kj} \tag{1.43}$$

Generally, the indicial format of the tensors is defined as follows:
Scalar: Tensor of rank zero.
Vector: Tensor of rank one with three components or V_i.
Dyadic: Tensor of rank two with nine components, or T_{ij}.
Triadic: Tensor of rank three with twenty-seven components, or C_{ijk}.
Tetradic: Tensor of rank four with eighty-one components or t_{ijkm}.

The Eigenvector and Eigenvalue

To explain the concept of an eigen vector and value, we express below a simple example:

$$\begin{pmatrix} 1 & 2 \\ 8 & 1 \end{pmatrix}\begin{pmatrix} 1 \\ 2 \end{pmatrix} = 5\begin{pmatrix} 1 \\ 2 \end{pmatrix} = \begin{pmatrix} 5 \\ 10 \end{pmatrix}$$

In this example, the eigenvector $\begin{pmatrix} 1 \\ 2 \end{pmatrix}$, is the vector that, when it is multiplied by a matrix, then the result, $\begin{pmatrix} 5 \\ 10 \end{pmatrix}$, it is the same as when the vector is multiplied by a scalar of 5. Therefore, we can formulate this example as:

$$\vec{y} = A\,.\,\vec{x} \equiv \lambda.\,\vec{x} \qquad \text{(a)}$$

where A is a matrix, x, y are vectors, and λ is a scalar. The above equation can be stated equivalently as:

$$y^{\rightarrow} = A\,.x^{\rightarrow} = \lambda.x^{\rightarrow} \equiv \lambda.\,I.x^{\rightarrow} \qquad \text{(b)}$$

$$A.x^{\rightarrow} - \lambda.\,I.x^{\rightarrow} = 0 \rightarrow (A - \lambda I)\,x^{\rightarrow} = 0 \qquad \text{(c)}$$

Equation c has a nonzero solution if and only if the determinant of the matrix $(A - \lambda I)$ is zero. Therefore, the eigenvalues of A are values of λ that satisfy the equation:

$$|A - \lambda I| = 0 \tag{1.44}$$

$$\rightarrow \left[\begin{pmatrix} A_{11} & A_{12} & A_{13} \\ A_{21} & A_{22} & A_{23} \\ A_{31} & A_{32} & A_{33} \end{pmatrix} - \lambda\begin{pmatrix} 1 & 0 & 0 \\ 0 & 1 & 0 \\ 0 & 0 & 1 \end{pmatrix}\right]\begin{bmatrix} x_1 \\ x_2 \\ x_3 \end{bmatrix} = \begin{bmatrix} 0 \\ 0 \\ 0 \end{bmatrix} \qquad \text{(d)}$$

$$\rightarrow \begin{pmatrix} A_{11} - \lambda & A_{12} & A_{13} \\ A_{21} & A_{22} - \lambda & A_{23} \\ A_{31} & A_{32} & A_{33} - \lambda \end{pmatrix}\begin{pmatrix} x_1 \\ x_2 \\ x_3 \end{pmatrix} = \begin{pmatrix} 0 \\ 0 \\ 0 \end{pmatrix} \qquad \text{(e)}$$

$$\begin{cases} (A_{11} - \lambda)x_1 + A_{12}x_2 + A_{13}x_3 = 0 \\ A_{21}x_1 + (A_{22} - \lambda)x_2 + A_{23}x_2 = 0 \\ A_{31}x_1 + A_{32}x_2 + (A_{33} - \lambda) = 0 \end{cases} \qquad \text{(f)}$$

$$\begin{vmatrix} A_{11} - \lambda & A_{12} & A_{13} \\ A_{21} & A_{22} - \lambda & A_{23} \\ A_{31} & A_{32} & A_{33} - \lambda \end{vmatrix} = 0 \qquad \text{(g)}$$

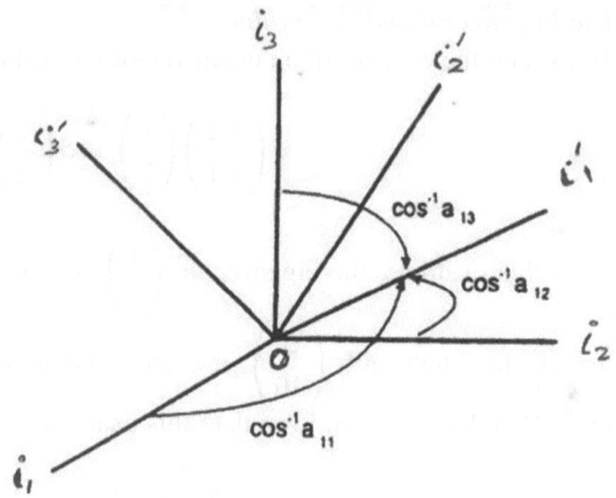

Fig. 1.3 The rotation of the Cartesian coordinate system and the related direction cosines

The determinant results in a polynomial equation of the third degree with its roots of λ_1, λ_2, λ_3 that refers to the eigenvalues of the matrix A. Each of the eigenvalues corresponds to an eigenvector using equation f.

Rotational Coordinate Transformations

In this section, we intend to study the rotation of the Cartesian coordinate system with a constant origin and obtain the new tensor components after the rotation of the coordinate system in the new one. See the Fig. 1.3, which shows the preliminary Cartesian coordinate system and the rotated one. $O_{i1,i2,i3}$, $O_{i'_1,i'_2,i'_3}$, The systems have a constant origin of O, and the direction cosines are also shown in the figure.

The pattern of changing direction cosines relative to i'_1 is based on the following:

$$a_{11} = \cos(i'_1, i_1)$$

$$a_{12} = \cos(i'_1, i_2)$$

$$a_{13} = \cos(i'_1, i_3)$$

In the general form of an indicial notation, we can write:

$$a_{ij} = \cos(i'_i, i_j) \tag{1.45}$$

To better understand the concept of the rotation of the Cartesian coordinate system, the following table is introduced, which includes nine direction cosines of unit vectors in the original and rotated coordinate systems:

		θ_1	θ_2	θ_3
		i_1	i_2	i_3
θ'_1	i'_1	a_{11}	a_{12}	a_{13}
θ'_2	i'_2	a_{21}	a_{22}	a_{33}
θ'_3	i'_3	a_{31}	a_{32}	a_{33}

We can use this table to calculate the unit vectors of the rotated coordinate system in terms of the original one:

$$i'_1 = a_{11}i_1 + a_{12}i_2 + a_{13}i_3 = a_{1j}.i_j$$

$$i'_2 = a_{21}i_1 + a_{22}i_2 + a_{23}i_3 = a_{2j}.i_j$$

$$i'_3 = a_{31}i_1 + a_{32}i_2 + a_{33}i_3 = a_{3j}.i_j$$

$$\rightarrow i'_i = a_{ij}i_j \tag{1.46}$$

Or:

$$\begin{bmatrix} i'_1 \\ i'_2 \\ i'_3 \end{bmatrix} = \begin{bmatrix} \mathrm{a}_{11}\ \mathrm{a}_{12}\ a_{13} \\ a_{21}\ a_{22}\ a_{23} \\ a_{31}\ a_{32}\ a_{33} \end{bmatrix} \begin{bmatrix} i_1 \\ i_2 \\ i_3 \end{bmatrix}$$

If we want to calculate the original unit vectors using the rotated ones, then:

$$[i'_i] = [a_{ij}][i_j] \rightarrow [i_i] = [a_{ij}]^{-1}.[i'_j] \tag{1.47}$$

We intend to prove $[a_{ij}]^{-1} = [a_{ij}]^T$ to rewrite relation (1.47):
From relation (1.46): $i'_i = a_{ij}i_j$. We write:

$$i'_i.i'_j = (a_{iq}.i_q).(i_{jn}.i_{\mathrm{m}}) \tag{1.48}$$

$$i'_i.i'_j = (a_{iq}.i_q).(i_{jm}.i_{\mathrm{m}}) = a_{iq}.a_{jm}.i_q.i_{\mathrm{m}} \tag{1.49}$$

It is interesting to know that all $i_{\mathrm{i},}i_{\mathrm{j}}$, i'_i, i'_j are unit vectors of the original and rotated Cartesian coordinate system, so relation (1.32) is still valid for them:

$$i_q.i_{\mathrm{m}} = \delta_{qm} \rightarrow i'_i : i'_j = a_{iq}.a_{jm}.\delta_{qm} \tag{1.50}$$

$$i'_i : i'_j = \delta_{ij} \rightarrow \left\{ \begin{array}{c} \delta_{ij} = a_{iq}.a_{jm}\delta_{qm} \\ a_{jm}.\delta_{qm} \equiv a_{jq} \end{array} \right\} \rightarrow \delta_{ij} = a_{iq}a_{jq}$$

$$\left\{ \begin{array}{c} \delta_{ij} = a_{iq}.a_{jm}\delta_{qm} \\ a_{jm}.\delta_{qm} \equiv a_{jq} \end{array} \right\} \rightarrow \delta_{ij} = a_{iq}a_{jq} \tag{1.51}$$

Relation (1.51) gives an important result to us:

$$\begin{aligned} \delta_{11} &= a_{11}.a_{11} + a_{12}.a_{12} + a_{13}.a_{13} \\ \delta_{12} &= a_{11}.a_{21} + a_{12}.a_{22} + a_{13}.a_{23} \\ \delta_{13} &= a_{11}.a_{31} + a_{12}.a_{32} + a_{13}.a_{33} \\ \delta_{22} &= a_{21}.a_{21} + a_{22}.a_{22} + a_{23}.a_{23} \end{aligned} \tag{1.52}$$

If we note relation (1.52), we can rewrite it in matrix form:

$$\underbrace{\begin{bmatrix} a_{11} & a_{12} & a_{13} \\ a_{21} & a_{22} & a_{23} \\ a_{31} & a_{32} & a_{33} \end{bmatrix}}_{\mathrm{A}} \underbrace{\begin{bmatrix} a_{11} & a_{21} & a_{31} \\ a_{12} & a_{22} & a_{32} \\ a_{13} & a_{23} & a_{33} \end{bmatrix}}_{\mathrm{A^T}} = \begin{bmatrix} \delta_{11} & \delta_{12} & \delta_{13} \\ \delta_{21} & \delta_{22} & \delta_{23} \\ \delta_{31} & \delta_{32} & \delta_{33} \end{bmatrix} \tag{1.53}$$

Relation (1.53) clearly shows that the transpose of matrix A is equal to its inverse:

$$A.A^T = I \rightarrow A^T = A^{-1}$$

Now we can rewrite relation (1.47) once again using 53:

$$[i_i] = [a_{ij}]^{-1}.[i'_j] = [a_{ij}]^{\mathrm{T}}.[i'_j]$$

$$\begin{bmatrix} i_1 \\ i_2 \\ i_3 \end{bmatrix} = \begin{bmatrix} a_{11} & a_{21} & a_{31} \\ a_{12} & a_{22} & a_{32} \\ a_{13} & a_{23} & a_{33} \end{bmatrix} \begin{bmatrix} i'_1 \\ i'_2 \\ i'_3 \end{bmatrix} \rightarrow i_i = a_{ji}.i'_j \tag{1.54}$$

It should be noted that a_{ij} is the transformation matrix which enables us to transform the original Cartesian coordinate system to another rotated one under any angle, and the components of this matrix are $a_{ij} = Cos(i'_i, i_j)a_{ij} = Cos(i'_i, i_j)$.

Now, we choose a vector of $\overrightarrow{V}$ which originally was in the (i_1, i_2, i_3), The first coordinate system, and is now transferred to a rotated coordinate system of $i'_1 i'_2 i'_3$ and is called V'_{i} but note that the size the vector is always unchanged due to this transformation:

$$\vec{V} = \mathrm{V}_i.i_i \equiv \mathrm{V}_1 i_1 + \mathrm{V}_2 i_2 + \mathrm{V}_3 i_3$$

$$\vec{V} = V'_{\mathrm{j}}.i'_{\mathrm{j}} = V'_1 i'_1 + V'_2 i'_2 + V'_3 i'_3$$

$$\left\{\begin{array}{c}\vec{V}=V'_{\mathrm{j}}.i'_{\mathrm{j}}=\mathrm{V}_i.i_i\\ i_i=a_{ji}.i'_j\end{array}\right\}\rightarrow \vec{V}=V_i.i_i=V_i\left(a_{ji}.i'_j\right)=V'_j.i'_j$$
$$=V_i\left(a_{ji}.i'_j\right)$$
$$=V'_j.i'_j \tag{1.55}$$

Note that a_{ji} is transpose of a_{ij} ansd it can be easily seen from relation (1.55) that:

$$V'_{\mathrm{j}}=a_{ji}.V_i \tag{1.56}$$

or:

$$\mathrm{V}'=[\mathrm{A}].\mathrm{V}=\mathrm{V}.[\mathrm{A}]^{\mathrm{T}}$$

Now, we multiple a_{jk} to both sides of relation (1.56) remembering that $a_{ij}.a_{jk}=\delta_{ik}$:

$$a_{jk}.V'_{\mathrm{j}}=a_{jk}.a_{ji}.V_i\rightarrow a_{jk}.V'_j=V_k$$
$$=\delta_{ik}.V_i=V_k \tag{1.57}$$

Or:

$$\vec{V}=\vec{V}'.[\mathrm{A}]\equiv[A]^T.\vec{V}' \tag{1.58}$$

In the next step, we consider two vectors of u_i and v_j and perform a Dyadic product:

$$u_{\mathrm{i}}=a_{qi}.u'_q$$

$$V_{\mathrm{j}}=a_{mj}.V'_m$$

$$u_{\mathrm{i}}\otimes \mathrm{V}_{\mathrm{j}}=\left(a_{\mathrm{qi}}.\mathrm{u}'_{\mathrm{q}}\right)\otimes\left(a_{\mathrm{mj}}.\mathrm{v}'_{\mathrm{m}}\right)$$
$$=a_{qi}.a_{mj}.\left(\mathrm{u}'_{\mathrm{q}}\otimes\mathrm{v}'_{\mathrm{m}}\right)$$

Based on the Dyadic product definition, we have:

$$\mathrm{u}'_{\mathrm{q}}\otimes\mathrm{v}'_{\mathrm{m}}=\mathrm{T}'_{\mathrm{qm}}$$

Therefore:

$$u_{\mathrm{i}}\otimes\mathrm{V}_{\mathrm{j}}=\mathrm{T}_{\mathrm{ij}}\equiv a_{qi}.a_{mj}.T'_{qm} \tag{1.59}$$

Or in matrix form:

$$T = A^T.T'.A \tag{1.60}$$

The inverse of relation (1.59) is expressed as:

$$T'_{ij} = a_{iq}.a_{jm}.T_{qm} \tag{1.61}$$

Or based on the matrix form:

$$T' = A.T.A^T \tag{1.62}$$

where A is the same as a_{ij} (Transformation tensor)

Relations (1.58), (1.60), and (1.62) have many applications in problems that require transferring a vector or a tensor from an original coordinate system to a rotated one.

Scalar Functions

A scalar function is a function that takes one or more input values and returns a single output value. See the example below:

$$\begin{aligned} f(x, y, z) &= x^2 + y + \frac{z}{2} \\ f(1, 2, 3) &= 1^2 + 2 + \frac{3}{2} = \frac{9}{2} \\ f(1, 1, 2) &= 1^2 + 1 + \frac{2}{2} = 3 \end{aligned} \tag{1.63}$$

The Gradient of a Scalar Function

The gradient of a function is defined to be a vector field. Generally, the gradient of a function can be found by applying the vector operator to the scalar function. (∇f (x, y, z)). This kind of vector field is known as the gradient vector field. The gradient denotes the direction of the greatest change of a scalar function.

$$\nabla f = \frac{\partial f}{\partial x} i + \frac{\partial f}{\partial y} j + \frac{\partial f}{\partial z} k$$

$$\nabla f = \frac{\partial f}{\partial x_1} i_1 + \frac{\partial f}{\partial x_2} i_2 + \frac{\partial f}{\partial x_3} i_3 = \frac{\partial f}{\partial x_m} . i_m \tag{1.64}$$

Divergence

Divergence is a vector operator, namely, Del, ∇, which gives the vector field's source quantity at each point. It measures the magnitude of a vector field, like a source or sink, and when it operates on a vector, the result is a scalar:

$$\nabla = \frac{\partial}{\partial x_1} i_1 + \frac{\partial}{\partial x_2} i_2 + \frac{\partial}{\partial x_3} i_3 (\text{Del})$$

$$\text{div}\ \overrightarrow{A} = \nabla.\overrightarrow{A} = \left(\frac{\partial}{\partial x_1}i_1 + \frac{\partial}{\partial x_2}i_2 + \frac{\partial}{\partial x_3}i_3\right).(A_1i_1 + A_2i_2 + A_3i_3)$$
$$\equiv \frac{\partial A_1}{\partial x_1} + \frac{\partial A_2}{\partial x_2} + \frac{\partial A_3}{\partial x_3} \quad (1.65)$$

$$\vec{A} = \mathrm{A_1i_1 + A_2i_2 + A_3i_3}$$

The divergence of vector A is the result of its dot product with the Del operator, which gives a scalar. As a physical example, consider a heated or cooled air where the air velocity is the velocity vector field at each point. When the air is heated, there will be an expansion in all directions, and the direction of the velocity vector field is outward. That means the divergence of the velocity vector field is positive in that region, and we call it a heat source. Vice versa, when the air is cooled, then there will be a contraction in all directions, and the velocity vector field direction is inward with a negative divergence and is called a sink that absorbs heat.

Curl

The curl of a vector is calculated by the cross product of a vector $\vec{A}$ and Del operator:

$$Curl\vec{A} = \nabla \times \vec{A} = \begin{vmatrix} i_1 & i_2 & i_3 \\ \frac{\partial}{\partial x_1} & \frac{\partial}{\partial x_2} & \frac{\partial}{\partial x_3} \\ A_1 & A_2 & A_3 \end{vmatrix} \quad (1.66)$$

Consider a vector field that represents a fluid velocity: The curl of the field at a point in a fluid is a measure of the rotation of the fluid. If there is no rotation of fluid anywhere, such a vector field is said to be irrotational. In a physical example, consider a full bath of water with a whirlpool inside it. If the whirligig starts to twist in a special point in the bath and we consider the water velocity as a vector field, then the rotation of the whirligig represents the curl of the water velocity vector field at the point. If the whirligig does not twist at a point, that means the curl of the water velocity vector is zero at that point.

Divergence Theorem

The theorem is applicable in future chapters and is expressed in a simple form as below: V represents a volume in a three-dimensional space with a smooth closed boundary S. Using the theorem, we can change the volume integral to a surface integral for a vector field like A by:

$$\int_V div\vec{A}.\mathrm{dv} = \int_s n.\vec{A}.ds \quad (1.67)$$

n is the outward-pointing unit normal at every point on the boundary.

The relation represents that the volume integral of the divergence of vector A is equal to the surface integral of the dot product of vector A and the unit normal n. In the rest of the chapter, we solved some representative problems which are very useful for a better understanding of the formulas explained and relations:

1. For tensor $T_{\mathrm{ij}} = \begin{bmatrix} 1 & 0 & 2 \\ 0 & 1 & 2 \\ 3 & 0 & 3 \end{bmatrix}$, calculate $T_{mn}.T_{mm}$, $T_{ij}.T_{ij}$, T_{ii}

Solution

In each case, we expand the indicial notation sentence and replace the equivalent values from the tensor:

$$T_{\mathrm{ij}} = T_{11} + \mathrm{T}_{22} + \mathrm{T}_{33} = 1 + 1 + 2 = 4$$

$$\begin{aligned} T_{\mathrm{ij}}.T_{\mathrm{ij}} &= T_{11}.T_{11} + T_{12}.T_{12} + T_{13}.T_{13} + T_{21}.T_{21} + T_{22}.T_{22} \\ &+ T_{23}.T_{23} + T_{31}.T_{31} + T_{32}.T_{32} + T_{33}.T_{33} \\ &= 1 \times 1 + 0 + 2 \times 2 + 0 + 1 + 2 \times 2 + 3 \times 3 + 0 + 3 \times 3 = 28 \end{aligned}$$

$$\begin{aligned} T_{mn}.\mathrm{T}_{mm} &\equiv (T_{11}T_{11} + \mathrm{T}_{21}T_{22} + T_{31}\mathrm{T}_{33}) + (T_{12}T_{11} + \mathrm{T}_{22}T_{22} + T_{32}\mathrm{T}_{33}) \\ &+ (T_{13}T_{11} + \mathrm{T}_{23}T_{22} + T_{33}\mathrm{T}_{33}) \\ &\equiv (1 \times 1 + 0 + 3 \times 3) + (0 + 1 + 0) + (3 \times 1 + 2 \times 1 + 3 \times 3) \\ &\equiv (1 + 9) + 1 + (3 + 2 + 9) = 24 \end{aligned}$$

2. Determine the result of the following indicial sentences.

(2.1) $$\begin{aligned} \delta_{ij}.\delta_{ij} &= \delta_{11}\delta_{11} + \delta_{12}\delta_{12} + \delta_{13}\delta_{13} + \delta_{21}\delta_{21} + \delta_{22}\delta_{22} \\ &+ \delta_{31}\delta_{31} + \delta_{32}\delta_{32} + \delta_{33}\delta_{33} \\ &= 1 \times 1 + 0 + 0 + 0 + 1 \times 1 + 0 + 0 + 1 \times 1 = 3 \end{aligned}$$

(2.2) $\delta_{ij}.\delta_{jk}.\delta_{ki} = \delta_{ij}.\delta_{jk} = \delta_{11}\delta_{11} + \delta_{22}\delta_{22} + \delta_{33}\delta_{33} = 1 \times 1 + 1 \times 1 + 1 \times 1 = 3$

Note that other elements are zero including $\delta_{13}.\delta_{31}$, $\delta_{23}.\delta_{32}$, $\delta_{12}.\delta_{21}$.

$$e_{ijk}.\delta_{kj} = e_{111}.\delta_{11} + e_{112}.\delta_{21} + e_{113}\delta_{31} + e_{121}\delta_{12} + e_{122}.\delta_{22} + \ldots + e_{133}.\delta_{33} + \ldots = 0$$

In the first sentence $e_{111} = 0$ and other Kronecker delta sentences like $\delta_{12}.\delta_{31}$ are zero. On the other hand, $e_{133} = e_{122} = 0$. In the end, other permutation symbols like $e_{222}\delta_{22}$, $e_{333}\delta_{33}$ are also zero, therefore the result is equal to zero.

$$\begin{aligned}\delta_{ik}.e_{ikm} &= \delta_{11}.e_{111} + \delta_{11}.e_{112} + \delta_{11}.e_{113} + \delta_{22}.e_{221}\\ &+ \delta_{22}.e_{222} + \delta_{22}.e_{223} + \delta_{33}.e_{331} + \delta_{33}.e_{332}\\ &+ \delta_{33}.e_{333} \equiv 0 + 0 + 0 + \ldots = 0\end{aligned}$$

$$\begin{aligned}&e_{ijk}.A_j.A_k = e_{123}.A_2A_3 + e_{213}.A_1A_3 + e_{312}.A_1A_2\\ &+ e_{132}.A_3A_2 + e_{231}.A_3A_1 + e_{321}.A_2A_1 =\\ &A_2A_3 - A_1A_3 + A_1A_2 - A_3A_2 + A_3A_1 - A_1A_2 = 0\end{aligned}$$

In problems (2.4) and (2.5), we have not written the permutation symbols with repeated indices like e_{111}, e_{122}, or e_{233} as they are zero.

$\nabla.(\nabla \times A)$

Suppose $\vec{A} = A_1 i_1 + A_2 i_2 + A_3 i_3$

$$\begin{aligned}\nabla \times A = \begin{vmatrix} i_1 & i_2 & i_3 \\ \frac{\partial}{\partial x_1} & \frac{\partial}{\partial x_2} & \frac{\partial}{\partial x_3} \\ A_1 & A_2 & A_3 \end{vmatrix} &= i_1\left(\frac{\partial A_3}{\partial x_2} - \frac{\partial A_2}{\partial x_3}\right)\\ &+ i_2\left(\frac{\partial A_1}{\partial x_3} - \frac{\partial A_3}{\partial x_1}\right) + i_3\left(\frac{\partial A_2}{\partial x_1} - \frac{\partial A_1}{\partial x_2}\right)\end{aligned}$$

$$\begin{aligned}\nabla.(\nabla \times A) &\equiv \left(\frac{\partial}{\partial x_1} i_1 + \frac{\partial}{\partial x_2} i_2 + \frac{\partial}{\partial x_1} i_3\right).\\ &\left[\left(\frac{\partial A_3}{\partial x_2} - \frac{\partial A_2}{\partial x_3}\right).i_1 + \left(\frac{\partial A_1}{\partial x_3} - \frac{\partial A_3}{\partial x_1}\right).i_2 + \left(\frac{\partial A_2}{\partial x_1} - \frac{\partial A_1}{\partial x_2}\right).i_3\right]\\ &\equiv \frac{\partial^2 A_3}{\partial x_1.\partial x_2} - \frac{\partial^2 A_2}{\partial x_1.\partial x_3} + \frac{\partial^2 A}{\partial x_2.\partial x_3} - \frac{\partial^2 A_3}{\partial x_1.\partial x_2}\\ &\frac{\partial^2 A_2}{\partial x_1.\partial x_3} - \frac{\partial^2 A_1}{\partial x_2.\partial x_3} \equiv 0\end{aligned}$$

3. First prove the following relation and then derive the magnitude of $e_{ijk}.e_{kji}$

$$e_{ijm}.e_{klm} = \delta_{ik}.\delta_{jl} - \delta_{il}.\delta_{jk}$$

Using relation (1.26), we substitute m for the last index and calculate the determinant by Sarus:

$$\mathrm{e_{ijm}.e_{klm}} \equiv \begin{vmatrix} \delta_{ik} & \delta_{il} & \delta_{im} \\ \delta_{jk} & \delta_{jl} & \delta_{jm} \\ \delta_{mk} & \delta_{ml} & \delta_{mm} \end{vmatrix}$$

$$\equiv \begin{vmatrix} \delta_{ik} & \delta_{il} & \delta_{im} \\ \delta_{jk} & \delta_{jl} & \delta_{jm} \\ \delta_{mk} & \delta_{ml} & \delta_{mm} \end{vmatrix} \begin{vmatrix} \delta_{ik} & \delta_{il} \\ \delta_{jk} & \delta_{jl} \\ \delta_{mk} & \delta_{ml} \end{vmatrix}$$

$$\equiv \delta_{ik}.\delta_{jl} + \delta_{il}.\delta_{jk} + \delta_{il}.\delta_{jk} - \delta_{il}.\delta_{jk} - \delta_{ik}.\delta_{jl} - \delta_{ik}.\delta_{jl}$$
$$= \delta_{il}.\delta_{jk} - \delta_{ik}.\delta_{jl}$$

$$\mathrm{e_{ijk}}.e_{kji} = \begin{vmatrix} \delta_{ik} & \delta_{ij} & \delta_{ii} \\ \delta_{jk} & \delta_{jj} & \delta_{ji} \\ \delta_{kk} & \delta_{kj} & \delta_{ki} \end{vmatrix} \equiv \begin{vmatrix} \delta_{ik} & \delta_{ij} & \delta_{ii} \\ \delta_{jk} & \delta_{jj} & \delta_{ji} \\ \delta_{kk} & \delta_{kj} & \delta_{ki} \end{vmatrix} \cdot \begin{bmatrix} \delta_{ik} & \delta_{ij} \\ \delta_{jk} & \delta_{jj} \\ \delta_{kk} & \delta_{kj} \end{bmatrix}$$

$$\equiv (\delta_{ik}.\delta_{jj}.\delta_{ki} + \delta_{ij}.\delta_{ji}.\delta_{kk} + \delta_{ii}.\delta_{jk}.\delta_{kj} - \delta_{ij}.\delta_{jk}.\delta_{ki} - \delta_{ik}.\delta_{ji}.\delta_{kj}$$
$$- \delta_{ii}.\delta_{jj}.\delta_{kk} \equiv 3\delta_{ik}.\delta_{ki} + 3\delta_{ij}.\delta_{ji} + 3\delta_{jk}.\delta_{kj} - \delta_{ij}.\delta_{jk}.\delta_{ki}$$
$$- \delta_{ik}.\delta_{ji}.\delta_{kj} - 3 \times 3 \times 3$$
$$\equiv 3\delta_{ii} + 3\delta_{ii} + 3\delta_{jj} - \delta_{ii} - \delta_{ii} - 27$$
$$\equiv 3 \times 3 + 3 \times 3 + 3 \times 3 - 3 - 3 - 27 \equiv -6$$

4. Consider the following vector field and derive the magnitude of Curl V and Div(V) at (1, 1, 0). $V = x^2 i_1 + z^2 i_2 + y^2 i_3$

$$div(V) = \frac{\partial(V_1)}{\partial x} + \frac{\partial(V_2)}{\partial y} + \frac{\partial(V_3)}{\partial z}$$

$$V = V_1 i_1 + V_1 i_1 + V_1 i_1 = x^2 i_1 + z^2 i_2 + y^2 i_3$$

$$div(V) = \frac{\partial(x^2)}{\partial x} + \frac{\partial(z^2)}{\partial y} + \frac{\partial(y^2)}{\partial z} = 2x + 0 + 0 = 2x$$

$$\text{point } (1, 1, 0) \to diV(V) = 2(1) = 2$$

$$Curl(V) = \nabla \times V \equiv \begin{vmatrix} i & j & k \\ \frac{\partial}{\partial x} & \frac{\partial}{\partial y} & \frac{\partial}{\partial z} \\ x^2 & z^2 & y^2 \end{vmatrix} = (2y - 2z)i + 0j + 0k = (2y - 2z)i$$

$$\text{point}(1, 1, 0) \to Curl(V) = [2(1) - 2(0)]i = 2i$$

5. For the vector $\vec{x} = x_i.e_i$ when $x^2 = x_1^2 + x_2^2 + x_3^2$, calculate the following items:

$grad\,x, div(x^n.\vec{x}),\ Curl(x^n.\vec{x})$

$$gradx = \nabla x = \frac{\partial x}{\partial x_1} i_1 + \frac{\partial x}{\partial x_2} i_2 + \frac{\partial x}{\partial x_3} i_3$$

$$x^2 = x_1^2 + x_2^2 + x_3^2$$

$$x = \sqrt{x_1^2 + x_2^2 + x_3^2} \rightarrow \frac{\partial x}{\partial x_1} = \frac{2x_1}{2\sqrt{x_1^2 + x_2^2 + x_3^2}},\ \frac{\partial x}{\partial x_2} = \frac{2x_2}{2\sqrt{x_1^2 + x_2^2 + x_3^2}},$$

$$\frac{\partial x}{\partial x_3} = \frac{2x_3}{2\sqrt{x_1^2 + x_2^2 + x_3^2}}$$

$$\rightarrow gradx = \nabla x = \frac{x_1 i_1 + x_2 i_2 + x_3 i_3}{\sqrt{x_1^2 + x_2^2 + x_3^2}}$$

$$\begin{aligned}
div(x^n.\vec{x}) &= diV(x^n.\mathrm{x}_1.\mathrm{i}_1 + x^n.\mathrm{x}_2.\mathrm{i}_2 + x^n.\mathrm{x}_3.\mathrm{i}_3 \\
&\equiv \frac{\partial(x^n.\mathrm{x}_1)}{\partial x_1} + \frac{\partial(x^n.\mathrm{x}_2)}{\partial x_2} + \frac{\partial(x^n.\mathrm{x}_3)}{\partial x_3} \\
&= n.x^{n-1}.\frac{\partial x}{\partial x_1}.x_1 + n.x^{n-1}.\frac{\partial x}{\partial x_2}.x_2 + n.x^{n-1}.\frac{\partial x}{\partial x_3}.x_3 + x^n + x^n + x^n \\
&= n.x^{n-1}.\left(\frac{2x_1}{2\sqrt{x_1^2 + x_2^2 + x_3^2}}\right).x_1 \\
&+ n.x^{n-1}\left(\frac{2x_2}{2\sqrt{x_1^2 + x_2^2 + x_3^2}}\right).x_2 + n.x^{n-1}.\left(\frac{2x_3}{2\sqrt{x_1^2 + x_2^2 + x_3^2}}\right).x_3 + 3x^n \\
&= n.x^{n-1}.\left(\frac{x_1^2 + x_2^2 + x_3^2}{\sqrt{x_1^2 + x_2^2 + x_3^2}}\right) + 3x^n = n.x^{n-1}\left(\frac{x^2}{x}\right) + 3x^n \\
&= nx^n + 3x^n = x^n(3 + n)
\end{aligned}$$

Note that: $(x = \sqrt{x_1^2 + x_2^2 + x_3^2})$

$$Curl\left(x^n.\vec{x}\right) = Curl\left(x^n x_1 i_1 + x^n x_2 i_2 + x^n x_3 i_3\right.$$

$$= \begin{vmatrix} i_1 & i_2 & i_3 \\ \frac{\partial}{\partial x_1} & \frac{\partial}{\partial x_2} & \frac{\partial}{\partial x_3} \\ x^n x_1 & x^n x_2 & x^n x_3 \end{vmatrix} = i_1\left(\frac{\partial x^n x_3}{\partial x_2} - \frac{\partial x^n x_2}{\partial x_3}\right)$$

$$+ i_2\left(\frac{\partial x^n x_1}{\partial x_3} - \frac{\partial x^n x_3}{\partial x_1}\right) + i_3\left(\frac{\partial x^n x_2}{\partial x_1} - \frac{\partial x^n x_1}{\partial x_2}\right)$$

$$\equiv \left[i_1\left(\frac{x_2 x_3 - x_3 x_2}{\sqrt{x_1^2 + x_2^2 + x_3^2}}\right) + i_2\left(\frac{x_1 x_3 - x_1 x_3}{\sqrt{x_1^2 + x_2^2 + x_3^2}}\right) + i_3\left(\frac{x_1 x_2 - x_1 x_2}{\sqrt{x_1^2 + x_2^2 + x_3^2}}\right)\right] n.x^{n-1} \equiv 0$$

6. The degree between two original and rotated Cartesian coordinates, $Ox_1x_2x_3$, $Ox'_1x'_2x'_3$ are given in the table below.
   (6.1) Obtain the transformation tensor, a_{ij}
   (6.2) For the plate with the equation: $x_1 + x_2 + x_3 = \frac{\sqrt{2}}{2}$, calculate the equation of the plate in transformed (rotated) coordinate.

	x_1	x_2	x_3
x'_1	45°	90°	45°
x'_2	60°	45°	120°
x'_3	120°	45°	60°

$$a_{ij} = \mathrm{Cos}\left(x'_i, x_j\right) \rightarrow$$

$$a_{11} = \mathrm{Cos}\left(x'_1, x_1\right)$$
$$= Cos45^\circ = \frac{\sqrt{2}}{2}$$

$$a_{12} = \mathrm{Cos}\left(x'_1, x_2\right)$$
$$= Cos90^\circ = 0$$

$$a_{13} = \mathrm{Cos}\left(x'_1, x_3\right)$$
$$= Cos45^\circ = \frac{\sqrt{2}}{2}$$

$$a_{21} = \mathrm{Cos}\left(x'_2, x_1\right)$$

$$= Cos60^{\circ} = \frac{1}{2}$$

$$a_{22} = \mathrm{Cos}\left(x_2', x_2\right)$$
$$= Cos45^{\circ} = \frac{\sqrt{2}}{2}$$

$$a_{23} = \mathrm{Cos}\left(x_2', x_3\right)$$
$$= Cos120^{\circ} = -\frac{1}{2}$$

$$a_{31} = \mathrm{Cos}\left(x_3', x_1\right)$$
$$= Cos120^{\circ} = -\frac{1}{2}$$

$$a_{32} = \mathrm{Cos}\left(x_3', x_2\right)$$
$$= Cos45^{\circ} = \frac{\sqrt{2}}{2}$$

$$a_{33} = \mathrm{Cos}\left(x_3', x_3\right)$$
$$= Cos45^{\circ} = \frac{1}{2}$$

$$a_{ij} = \begin{bmatrix} \frac{\sqrt{2}}{2} & 0 & \frac{\sqrt{2}}{2} \\ \frac{1}{2} & \frac{\sqrt{2}}{2} & -\frac{1}{2} \\ -\frac{1}{2} & \frac{\sqrt{2}}{2} & \frac{1}{2} \end{bmatrix}$$

Now, we can calculate the transformed equation of the original plate with the normal unit.

$\vec{n}(1, 1, 1)$ using relation (1.56):

$$\overrightarrow{n'} = \left[a_{ij}\right]\vec{n} = \begin{bmatrix} \frac{\sqrt{2}}{2} & 0 & \frac{\sqrt{2}}{2} \\ \frac{1}{2} & \frac{\sqrt{2}}{2} & -\frac{1}{2} \\ -\frac{1}{2} & \frac{\sqrt{2}}{2} & \frac{1}{2} \end{bmatrix} \begin{bmatrix} 1 \\ 1 \\ 1 \end{bmatrix} = \begin{bmatrix} \frac{\sqrt{2}}{2} \\ \frac{\sqrt{2}}{2} \\ \frac{\sqrt{2}}{2} \end{bmatrix}$$

$$\overrightarrow{n'} = \left(\sqrt{2}, \frac{\sqrt{2}}{2}, \frac{\sqrt{2}}{2}\right)$$

Assume point A $\left(\frac{\sqrt{2}}{2}, 0, 0\right)$ on the original plate that we are going to transform it into the new rotated coordinate using the transformation tensor:

$$\begin{bmatrix} a'_1 \\ a'_2 \\ a'_3 \end{bmatrix} = \begin{bmatrix} \frac{\sqrt{2}}{2} & 0 & \frac{\sqrt{2}}{2} \\ \frac{1}{2} & \frac{\sqrt{2}}{2} & -\frac{1}{2} \\ -\frac{1}{2} & \frac{\sqrt{2}}{2} & \frac{1}{2} \end{bmatrix} \begin{bmatrix} \frac{\sqrt{2}}{2} \\ 0 \\ 0 \end{bmatrix} = \begin{bmatrix} \frac{1}{2} \\ \frac{\sqrt{2}}{4} \\ -\frac{\sqrt{2}}{4} \end{bmatrix}$$

Now, we obtain the new rotated plate equation by $a'\left(\frac{1}{2}, \frac{\sqrt{2}}{4}, \frac{-\sqrt{2}}{4}\right), \overrightarrow{n'}$: $a'\left(\frac{1}{2}, \frac{\sqrt{2}}{4}, \frac{-\sqrt{2}}{4}\right), \overrightarrow{n'}$.

The equation of rotated plate: $a'\mathrm{x} + \mathrm{b}'\mathrm{y} + \mathrm{c}'\mathrm{z} = \mathrm{d}'$

$$n' = \left(\sqrt{2}, \frac{\sqrt{2}}{2}, \frac{\sqrt{2}}{2}\right) = (a', b', c')$$

$$\rightarrow \sqrt{2}\mathrm{x} + \frac{\sqrt{2}}{2}\mathrm{y} + \frac{\sqrt{2}}{2}\mathrm{z} = \mathrm{d}'$$

$$a'\left(\frac{1}{2}, \frac{\sqrt{2}}{4}, \frac{-\sqrt{2}}{4}\right) \rightarrow \sqrt{2}\left(\frac{1}{2}\right) + \frac{\sqrt{2}}{2} \times \frac{\sqrt{2}}{4} + \frac{\sqrt{2}}{2}\left(-\frac{\sqrt{2}}{4}\right) = \mathrm{d}'$$

$$d' = \frac{\sqrt{2}}{2}$$

Therefore, the equation of the original plate in the rotated coordinate is: $2x + y + z = 1$ or $2x + y + z = 12x + y + z = 1$.

7- For the matrix $B_{\mathrm{ij}} = \begin{bmatrix} 17 & 0 & 0 \\ 0 & -23 & 28 \\ 0 & 28 & 10 \end{bmatrix}$ Determine the eigenvalues and related eigenvectors.

$$|B_{ij} - \lambda\delta_{ij}| = 0 \rightarrow \left| \begin{bmatrix} 17 & 0 & 0 \\ 0 & -23 & 28 \\ 0 & 28 & 10 \end{bmatrix} - \lambda \begin{bmatrix} 1 & 0 & 0 \\ 0 & 1 & 0 \\ 0 & 0 & 1 \end{bmatrix} \right| = 0$$

$$\rightarrow \begin{vmatrix} 17\lambda & 0 & 0 \\ 0 & -23\lambda & 28 \\ 0 & 28 & 10-\lambda \end{vmatrix} = 0 \rightarrow$$

$$\rightarrow (17-\lambda)\left[(-23-\lambda)(10-\lambda) - 28^2(17-\lambda\right] = 0$$

$$\rightarrow (17-\lambda)\left(\lambda^2 + 13\lambda - 1014\right) = 0 \rightarrow \begin{cases} \lambda_1 = 17 \\ \lambda_2 = 26 \\ \lambda_3 = -39 \end{cases}$$

To obtain the eigen vectors, we should use the expansion of $\left(B_{ij} - \lambda\delta_{ij}\right) = 0$ and note that $n_1^2 + n_2^2 + n_3^2 = 1 n_1^2 + n_2^2 + n_3^2 = 1$

$$\lambda_1 = 17 \rightarrow \begin{cases} -40n_2 + 28n_3 = 0 \\ 28n_2 + (-7)n_3 = 0 \end{cases} \rightarrow n_2 = n_3 = 0$$

$$n_1^2 + n_2^2 + n_3^2 = 1$$

$$n_1 = \pm 1$$

$$n_1^{(1)} = \pm \mathrm{i}$$

$$\lambda_2 = 26 \rightarrow \begin{cases} n_1 = 0 \\ n_2 = \frac{4}{7}n_3 \\ n_1^2 + n_2^2 + n_3^2 = 1 \rightarrow \begin{cases} n_3 = \frac{\pm 7}{\sqrt{65}} \\ n_2 = \frac{\pm 4}{\sqrt{65}} \end{cases} \end{cases}$$

$$n_2^{(2)} = \frac{\pm 4}{\sqrt{65}} j \pm \frac{7}{\sqrt{65}} k$$

$$\lambda_3 = 39 \rightarrow \begin{cases} n_1 = 0 \\ n_2 = \frac{-4}{7}n_3 \\ n_1^2 + n_2^2 + n_3^2 = 1 \rightarrow \begin{cases} n_3 = \frac{\pm 4}{\sqrt{65}} \\ n_2 = \frac{\mp 7}{\sqrt{65}} \end{cases} \end{cases}$$

$$n_2^{(3)} = \frac{\mp 7}{\sqrt{65}} j \pm \frac{4}{\sqrt{65}} k$$

Stress

2

In continuum mechanics, stress is a physical quantity that represents internal forces exerted among neighboring particles inside a continuum material. For example, when a rod is hung under its weight vertically, then every particle of the rod exerts a compressing force on the particle below, and all particles are under compression in four directions. On the other hand, strain is a concept for the magnitude of material deformation. The stress of either volumetric or surface ones leads to strain in a continuum material. The volumetric stress is induced by gravity, but the surface stress is induced by surface forces, friction, and external pressure. The stress can also be induced by temperature variations, chemical reactions, and electromagnetic fields. The relation between the mechanical stresses (like elastic or viscous), deformation, and rate of deformation is a complex material, although a linear approximation is generally precise for small strains.

Cauchy Stress Tensor

In continuum mechanics, the Cauchy stress tensor or true stress tensor, which we call simply the stress tensor from now on, is a (3×3) second-order tensor that consists of nine components and is shown as σ_{ij}. This tensor presents the exact stress situation of each point in a continuum material after deformation, change of configuration, and displacement. The tensor correlates the stress vector of T or $\left(T^{(n)}\right)$ to the normal vector of each surface of the cubic element that is shown in Fig. 2.1.

In the figure, each of the six surfaces has a normal vector. The cube is under equilibrium, and the stress vectors, including $T^{(e_1)}$, $T^{(e_2)}$, $T^{(e_3)}$, shown. If the normal vectors of these surfaces are expressed as n_1, n_2, n_3 then nine components of the stress tensor are calculated by:

K. Hassani and F. Azimifar, *Introduction to Continuum Mechanics*, Synthesis Lectures on Mechanical Engineering, https://doi.org/10.1007/978-3-032-05076-2_2

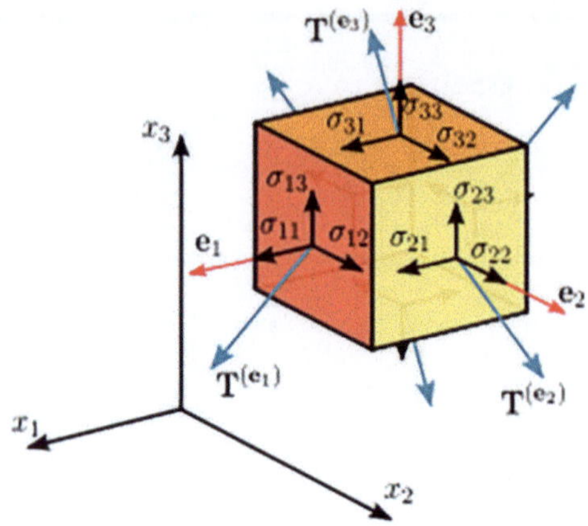

Fig. 2.1 The existing Cauchy stresses on the surfaces of the cubic element

$$T^{(n)} = n.\sigma \quad \text{or } T_j^{(n)} = \sigma_{ij}.n_i \tag{2.1}$$

$$\sigma_{ij} = \begin{bmatrix} \sigma_{11} & \sigma_{12} & \sigma_{13} \\ \sigma_{21} & \sigma_{22} & \sigma_{23} \\ \sigma_{31} & \sigma_{32} & \sigma_{33} \end{bmatrix} \tag{2.2}$$

The Concept of Cauchy Stress Tensor

Figure 2.2 shows two divided sections of a continuous body where both parts are in equilibrium, and internal forces and moments are shown in resultant vectors. The internal stress vector is also defined by $T^{(n)}$ as shown in the figure, where it acts on the surface S with the normal vector n. If P is a point inside the surface S, then we can define the exerted forces on the point as surface and volumetric forces.

$$\vec{F} = F_S + F_b \tag{2.3}$$

Fs → The surface force

Fb → The volumetric force

It should be noted that there is a moment ΔM On the surface, along with the forces mentioned. Now, we should focus on a very small surface that surrounds point P, and we call it ΔS with normal vector n.

The distribution of the force on point P consists of ΔF and surface moment ΔM. Now we can write:

$$\Delta F = T^{(n)}.\Delta s \tag{2.4}$$

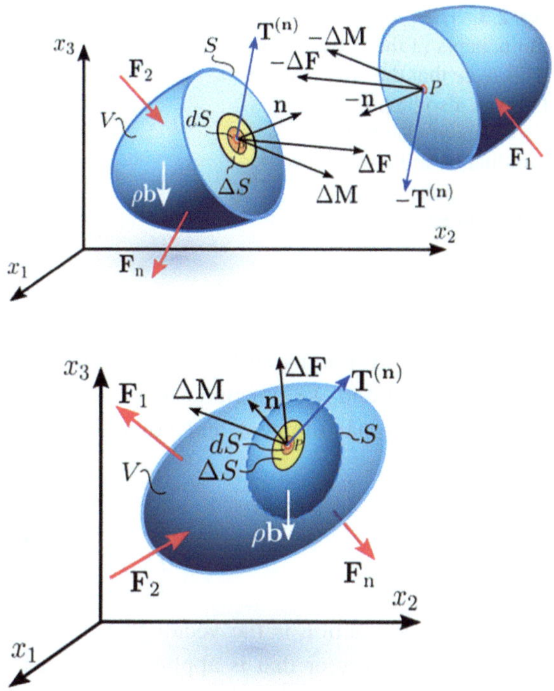

Fig. 2.2 The two cross sections of a continuous material with related internal forces and moments

$T^{(n)} \rightarrow$ Average surface tension

If ΔS is considered very small, near zero value, then the ratio of $\frac{\Delta F}{\Delta s}$ is changed to $\frac{dF}{ds}$ and two other events happened when $(\Delta S \rightarrow 0)$ as follows:

$$\Delta M = \Delta F_e.e \tag{2.5}$$

As we know, every couple or moment can be replaced by a couple of equal forces that have a distance e:

$$\Delta M = \Delta F_e.e$$

Now, if $\Delta s \rightarrow 0$ then e approaches zero also, which results $\Delta M = 0$.

Most of the time, the volumetric forces are only based on the gravitational force. Therefore, we can rewrite the volumetric force relation (2.3) as:

$$\vec{F}_b = m\vec{g} = \varrho V.\vec{g} \tag{2.6}$$

The volume of the surrounding material around point P is defined by:

$$\Delta V = \Delta S.t$$

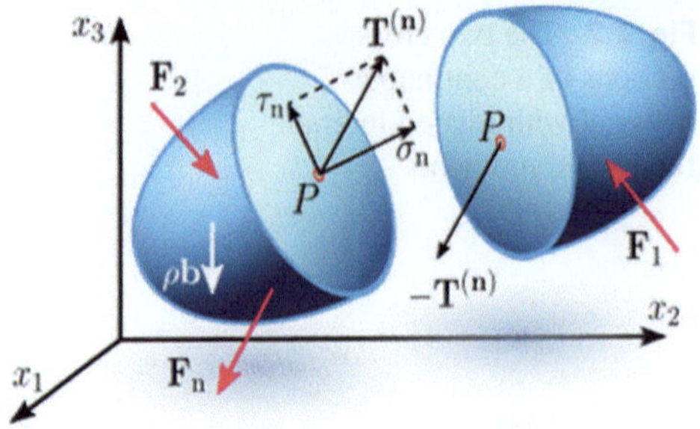

Fig. 2.3 The projected components of the stress vector $T^{(n)}$ *Note* that the stress vector $T^{(n)}$ can also be projected in directions of Cartesian coordinates and always has three components in x_1, x_2, and x_3 directions

where is the surrounding surface, and t is the thickness of the surface. Now, if $\Delta S \to 0$ then $\Delta V \to 0$ and we conclude that there is no volumetric force also. Therefore, we neglect the internal moment and volumetric forces during the study of Cauchy stress theory:

$$\left.\begin{array}{r}\Delta\vec{F}_b = \varrho\Delta V.\vec{g} \\ \Delta S \to 0\Delta\Delta V \to 0\end{array}\right\} \to \Delta\vec{F}_b = 0$$

The surface traction is the vector that is extracted by the ratio $\frac{dF}{dS}$ when if $\Delta S \to 0$. Thus, the stress vector that is applied on point P inside a plane with normal vector n is calculated by:

$$\begin{aligned} T^{(n)} &= T_i^{(n)}.e_i \\ T_i^{(n)} &= \lim_{\Delta s \to \cdot} \frac{\Delta F_i}{\Delta s} = \frac{dF_i}{ds} \end{aligned} \tag{2.7}$$

Relation (2.7) expresses that the stress vector depends on its position in the material and direction of the plane that it acts on. There is an important point that says the stress tensor is not always in the same direction as the normal vector n and shall be broken down into two perpendicular components as shown in Fig. 2.3.

$$\begin{aligned} (\text{Normal stress component})\sigma_n &= \lim_{\Delta s \to \cdot} \frac{\Delta F_n}{\Delta s} = \frac{dF_n}{ds} \\ (\text{Tangentialr shar component})\tau_n &= \lim_{\Delta s \to \cdot} \frac{\Delta F_s}{\Delta s} = \frac{dF_s}{ds} \end{aligned} \tag{2.8}$$

dF_n is the normal component of force dF, which acts on the ds surface and dF_s is the tangential component as well.

Cauchy Stress Theory

We consider a tetrahedron with three faces oriented in the coordinate planes, and area dA oriented in an arbitrary direction with a normal vector n. The stress vector acting on

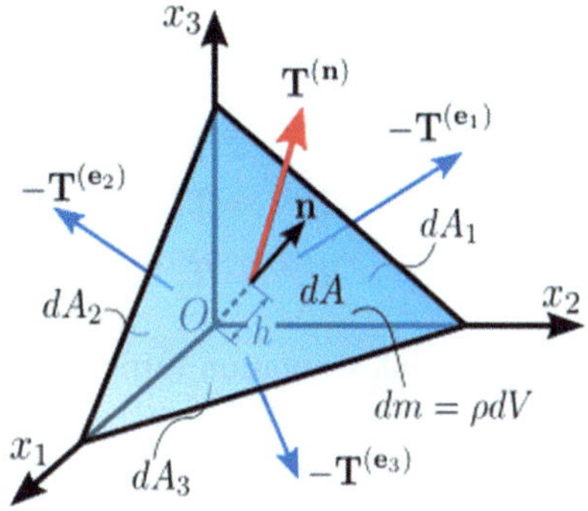

Fig. 2.4 Cauchy tetrahedron

the plane ABC is $T^{(n)}$ and $T^{(e_1)}, T^{(e_2)}, T^{(e_3)}$, or we call them from now $T^{(i_1)}, T^{(i_2)}, T^{(i_3)}$ Stress vectors are stress vectors that act on the tetrahedron faces. These vectors are components of the stress tensor σ_{ij}. Now, we write the second Newton law equations for the whole tetrahedron where a is the tetrahedron acceleration, h is the height of the tetrahedron, and ϱ is its density (Fig. 2.4).

$$T^{(n)}.dA - T^{(e_1)}.dA_1 - T^{(e_2)}.dA_2 - T^{(e_3)}.dA_3 = \varrho\left(\frac{h}{3}.dA\right)a \tag{2.9}$$

On the other hand:

$$\begin{aligned} dA_1 &= (n.i_1)dA = n_1 dA \\ dA_2 &= (n.i_2)dA = n_2 dA \\ dA_3 &= (n.i_3)dA = n_3 dAJ; \end{aligned} \tag{2.10}$$

Substituting relation (2.10) into (2.9):

$$T^{(n)} - T^{(i_1)}.n_1 - T^{(i_2)}.n_2 - T^{(i_3)}.n_3 = \sigma\left(\frac{h}{3}\right)a \tag{2.11}$$

In the limit case that surrounding material around o has a very small area, the height value approaches zero therefore:

$$T^{(n)} - T^{(i_1)}.n_1 - T^{(i_2)}.n_2 - T^{(i_3)}.n_3 = 0 \tag{2.12}$$

Let go back to Fig. 2.1 again to see that the cube faces are perpendicular to the Cartesian coordinates and related acting stress vectors are again $T^{(i_1)}, T^{(i_2)}, T^{(i_3)}$ and they can be projected to one normal and two tangential components in each face. In a special case, the normal vector of each face including n_1, n_2, n_3 is in the orientation of the coordinate's axes x_1, x_2, x_3. For example, for x_1 direction, we can call σ_{11} as normal stress and σ_{12}, σ_{13} as tangential or shear stresses of $T^{(e_1)}$ which act on the plane perpendicular to x_1. Now we rewrite stress vector components based on normal and shear stresses such as:

$$\begin{aligned}
T^{(i_1)} &= T_1^{(i_1)}.i_1 + T_2^{(i_1)}.i_2 + T_3^{(i_1)}.i_3 = \sigma_{11}i_1 + \sigma_{12}i_2 + \sigma_{13}i_3 \\
T^{(i_2)} &= T_1^{(i_2)}.i_1 + T_2^{(i_2)}.i_2 + T_3^{(i_2)}.i_3 = \sigma_{21}i_1 + \sigma_{22}i_2 + \sigma_{23}i_3 \\
T^{(i_3)} &= T_1^{(i_3)}.e_1 + T_2^{(i_3)}.i_2 + T_3^{(i_3)}.e_3 = \sigma_{31}i_1 + \sigma_{32}i_2 + \sigma_{33}i_3
\end{aligned} \tag{2.13}$$

Or in index notation format:

$$T^{(i_i)} = T_j^{(i_i)}.i_j = \sigma_{ij}i_j \tag{2.14}$$

Relation (2.13) can be written as follows:

$$T^{(n)} = T^{(i_1)}.n_1 + T^{(i_2)}.n_2 + T^{(i_3)}.n_3 = T^{(i_i)}.n_i \equiv (\sigma_{ij}.i_j)n_i \tag{2.15}$$

Equivalently:

$$T_j^{(n)} = \sigma_{ij}n_i \tag{2.16}$$

In fact, relation (2.16) is the same as 14, assuming the normal vectors n_1n_2, n_3 are in the orientation of Cartesian coordinates x_1x_1, x_1. Relation (2.16) is the main Cauchy stress relation and is applied in many problems.

The Transformation Rule of Stress Tensor

Figure 2.5 presents that if the tensor stress σ_{ij} is transformed by the transformation tensor a_{ij} Then the transformed stress tensor σ'_{ij} in new coordinates is expressed as relation (2.17). We have seen this rule previously in the mathematics chapter based on relation (1.62):

$$\sigma'_{ij} = a_{im}.a_{jn}.\sigma_{mn} or\ \sigma' = A\sigma.A^T \tag{2.17}$$

If we write relation (2.17) in matrix format:

$$\begin{bmatrix} \sigma'_{11} & \sigma'_{12} & \sigma'_{13} \\ \sigma'_{21} & \sigma'_{22} & \sigma'_{23} \\ \sigma'_{31} & \sigma'_{32} & \sigma'_{33} \end{bmatrix} = \begin{bmatrix} a_{11} & a_{12} & a_{13} \\ a_{21} & a_{22} & a_{23} \\ a_{31} & a_{32} & a_{33} \end{bmatrix} \begin{bmatrix} \sigma_{11} & \sigma_{12} & \sigma_{13} \\ \sigma_{21} & \sigma_{22} & \sigma_{23} \\ \sigma_{31} & \sigma_{32} & \sigma_{33} \end{bmatrix} \cdot \begin{bmatrix} a_{11} & a_{21} & a_{31} \\ a_{12} & a_{22} & a_{32} \\ a_{13} & a_{23} & a_{33} \end{bmatrix} \tag{2.18}$$

We calculate σ'_{11} using relation (2.18) but other components can be calculated as practice by readers:

$$\begin{aligned}
\sigma'_{11} =& a_{11}^2.\sigma_{11} + a_{12}^2.\sigma_{22} + a_{13}^2.\sigma_{33} + 2a_{11}.a_{12}\sigma_{12} \\
&+ 2a_{11}.a_{13}\sigma_{13} + 2a_{12}.a_{13}\sigma_{23}
\end{aligned} \tag{2.19}$$

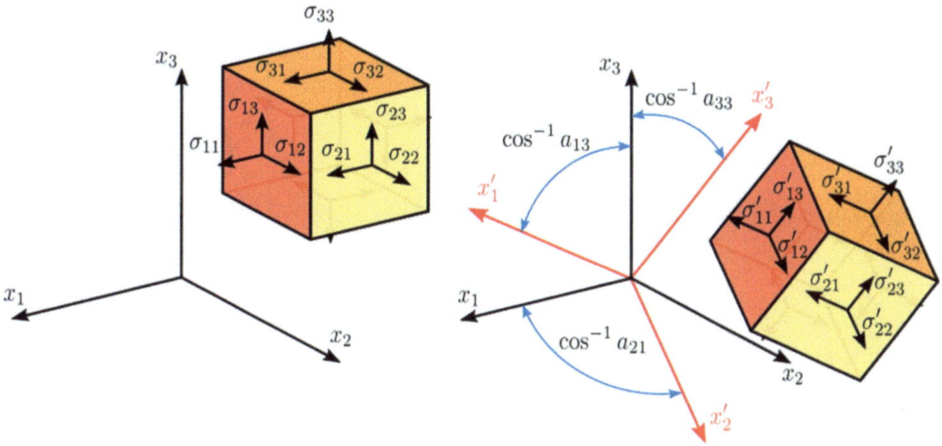

Fig. 2.5 Transformation of stress tensor

Fig. 2.6 The projection of a stress vector oriented in the normal vector direction: normal stress component

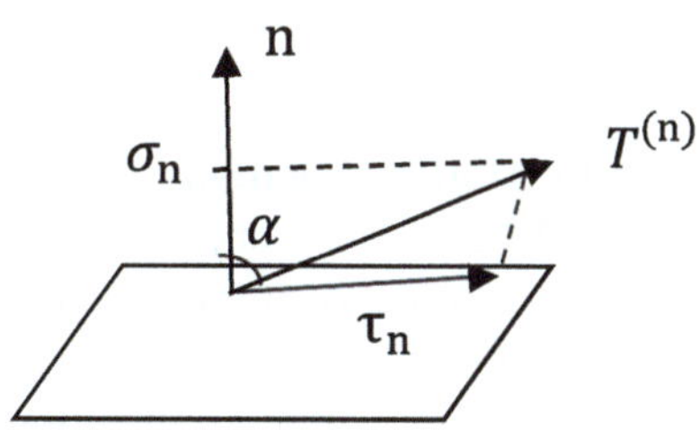

Normal and Shear Stresses

Consider a stress vector $T^{(n)}$ which is applied on a point in a plane with normal vector n. The magnitude of the normal stress component σ_n is calculated by the dot product vectors $T^{(n)}$ and the normal vector of the plane, which is n:

$$\sigma_n = \left|T^{(n)}\right|.cos\alpha \equiv \left|T^{(n)}\right|.|n|.cos\alpha \equiv \vec{T}^{(n)}.\vec{n} \tag{2.20}$$

Figure 2.6 shows that σ_n is the projection of the vector $T^{(n)}$ in normal vector n orientation or:

$$\begin{aligned}\sigma_n = T^{(n)}.n \equiv T_i^{(n)}.n_i = \sigma_{ij}.n_i.n_j \\ = \sigma_{11}.n_1^2 + \sigma_{22}.n_2^2 + \sigma_{33}.n_3^2 + 2\sigma_{12}n_1n_2 \\ + 2\sigma_{13}n_1n_3 + 2\sigma_{23}n_2n_3\end{aligned} \tag{2.21}$$

$\sigma_{23} = \sigma_{32}$ Note that: $\sigma_{13} = \sigma_{31}$, $\sigma_{12} = \sigma_{21}$

Tangential or shear component of the stress tensor τ_n which is perpendicular to the normal vector n is calculated by:

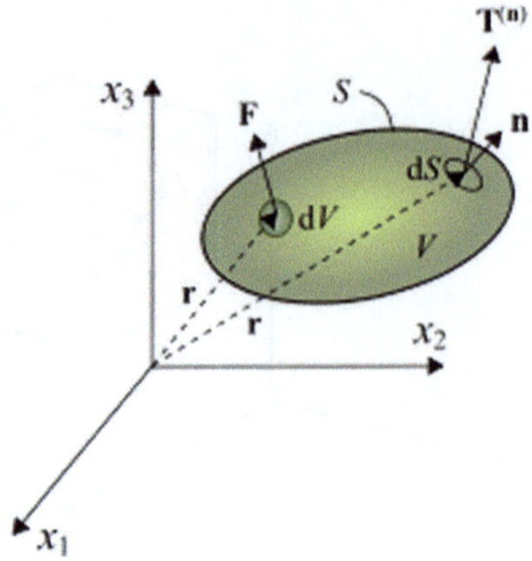

Fig. 2.7 A continuum body with volume V and surface S

$$\tau_n = \sqrt{\left(T^{(n)}\right)^2 - \sigma_n^2} = \sqrt{T_i^{(n)}.T_i^{(n)} - \sigma_n^2}$$

$$\left(T^{(n)}\right)^2 = T_i^{(n)}.T_i^{(n)} = \left(\sigma_{ij}.n_j\right)\left(\sigma_{ik}.n_k\right) \equiv \sigma_{ij}.\sigma_{ik}.n_j.n_k \tag{2.22}$$

Cauchy Motion Equations

Figure 2.7 shows a continuum body with volume V and surface S that has a stress vector $T_i^{(n)}$ acts on one of its points. The vector is the resultant of all surface forces per unit area, and Fi is the resultant of volumetric forces that act on volume V. We assume that the body is under equilibrium; therefore, the summation of the surface and volumetric forces is zero based on Newton's second law:

$$\int_S T_i^{(n)}.d_s + \int_V F_i.dv = 0 \tag{2.23}$$

Noting Again to $T_i^{(n)} = \sigma_{ji}.n_j$

$$\int_S T_i^{(n)}.d_s + \int_V F_i.dv = 0 \tag{2.24}$$

We now use the divergence theorem, which converts the volume integral to a surface integral. We will convert the surface integral of $T_i^{(n)}.d_s$ to a volume integral based on the theorem. From now on, we will also show differential relations as $\sigma_{ji,j} = \frac{\partial \sigma_{ji}}{\partial x_j}$:

$$\int_V \sigma_{ji,j} d_v + \int_V F_i.dv = 0 \tag{2.25}$$

$$\int_V (\sigma_{ji,j} + F_i) dv = 0 \tag{2.26}$$

$$\rightarrow \sigma_{ji,j} + F_i = 0$$

Relation (2.26) is the Cauchy motion equations, and it should also be noted that $\sigma_{ij} = \sigma_{ji}$ because the stress tensor is symmetrical. The next relations are derived from the expansion of relation (2.26):

$$\sigma_{ji,j} + F_i = 0 \rightarrow \begin{cases} \frac{\partial\sigma_{11}}{\partial x_1} + \frac{\partial\sigma_{12}}{\partial x_2} + \frac{\partial\sigma_{13}}{\partial x_3} + F_1 = 0 \\ \frac{\partial\sigma_{21}}{\partial x_1} + \frac{\partial\sigma_{22}}{\partial x_2} + \frac{\partial\sigma_{23}}{\partial x_3} + F_2 = 0 \\ \frac{\partial\sigma_{31}}{\partial x_1} + \frac{\partial\sigma_{32}}{\partial x_2} + \frac{\partial\sigma_{33}}{\partial x_3} + F_3 = 0 \end{cases} \tag{2.27}$$

Relation (2.27) is normally used to investigate the equilibrium situation of a continuum body.

Principal Stresses and Stress Invariants

In this section, we discuss the concept of principal stress and the related stress invariant. In the beginning, it should be said that the principal stresses and their eigenvectors are calculated using eigenvalues and eigenvectors relations of Chap. 1. In every point of a continuum environment that is under stress, there are at least three planes called principal planes. The property of these planes is that the acting stress vector is in the orientation of the normal vector of the plane. Therefore, there is no tangential or shear stress component in these planes, and normal stress is only found, which is called principal stress. The components of the stress tensor always depend on the directions of the coordinate system, but the tensor itself is a physical concept and is independent of any coordinate system that is used for defining it. Therefore, there are stress invariants in any stress tensor that do not depend on the coordinate system. To explain it better, consider a vector in Cartesian coordinates where its components depend on the direction of the coordinates, but the size of the vector is a scalar that does not depend on the coordinates and is always constant. Similarly, any stress or strain tensor is related to invariant quantities, and one of the invariant quantities includes principal stresses, which are eigenvalues of the stress tensor. All three vectors, which are in the orientation of the normal vectors, are called eigenvalues of the principal stresses (principal directions). Thus, we can define the stress vector when it is in the orientation of the normal vector of the plane $\overrightarrow{(n)}$:

$$T^{(n)} = \lambda n = \sigma_n . n \tag{2.28}$$

Relation (2.28) should be remembered from the last chapter when we explained the concept of the eigenvalue of the tensors. In this case, λ is the principal stress, and is the normal component of the stress vector. Knowing that $n_i = \delta_{ij} n_j$ and $T_i^{(n)} = \sigma_{ij} n_j$, we write:

$$\begin{aligned} &T_i^{(n)} = \lambda n_i \\ &\sigma_{ij} n_j = \lambda n_i \rightarrow \sigma_{ij} n_j - \lambda n_i = 0 \rightarrow \left(\sigma_{ij} - \lambda\delta_{ij}\right) n_j = 0 \end{aligned} \tag{2.29}$$

$$\left|\sigma_{ij} - \lambda\delta_{ij}\right| = \begin{vmatrix} \sigma_{11} - \lambda & \sigma_{12} & \sigma_{13} \\ \sigma_{21} & \sigma_{22} - \lambda & \sigma_{23} \\ \sigma_{31} & \sigma_{32} & \sigma_{33} - \lambda \end{vmatrix} = 0$$
$$\equiv -\lambda^3 + I_1\lambda^2 - I_2\lambda + I_3 = 0 \tag{2.30}$$

where:

$$I_1 = \sigma_{11} + \sigma_{22} + \sigma_{33} = \sigma_{KK}$$
$$I_2 = \begin{vmatrix} \sigma_{22} & \sigma_{23} \\ \sigma_{32} & \sigma_{33} \end{vmatrix} + \begin{vmatrix} \sigma_{11} & \sigma_{13} \\ \sigma_{31} & \sigma_{33} \end{vmatrix} + \begin{vmatrix} \sigma_{11} & \sigma_{12} \\ \sigma_{21} & \sigma_{22} \end{vmatrix}$$
$$= \sigma_{11}.\sigma_{22} + \sigma_{22}.\sigma_{33} + \sigma_{11}.\sigma_{33} - \sigma_{12}^2 - \sigma_{23}^2 - \sigma_{31}^2$$
$$= \frac{1}{2}\left(\sigma_{ii}\sigma_{jj} - \sigma_{ij}.\sigma_{ji}\right) \tag{2.31}$$

$$I_3 = \det\left(\sigma_{ij}\right) = \sigma_{11}.\sigma_{22}.\sigma_{33} + 2\sigma_{12}.\sigma_{23}.\sigma_{31} - \sigma_{12}^2\sigma_{33} - \sigma_{23}^2.\sigma_{11} - \sigma_{31}^2\sigma_{33}$$

Relation (2.30) has three real roots, which indicates three principal stresses of the stress tensor. The eigenvectors relating to these eigenvalues, λ_1, $\lambda_2\lambda_3$ principal stresses can be derived by relation (2.29). Principal stress and stress invariants are defined by the following relations:

$$\sigma_{ij} = \begin{vmatrix} \sigma_1 & 0 & 0 \\ 0 & \sigma_2 & 0 \\ 0 & 0 & \sigma_3 \end{vmatrix}$$
$$\begin{cases} I_1 = \sigma_1 + \sigma_2 + \sigma_3 \\ I_2 = \sigma_1\sigma_2 + \sigma_2\sigma_3 + \sigma_3\sigma_1 \\ I_3 = \sigma_1\sigma_2\sigma_3 \end{cases} \tag{2.32}$$

Minimum and Maximum Values of Shear Stresses

There is an interesting point in relation (2.21) that allows us to rewrite it based on principal stresses because the principal planes are the same as any other planes that can pass through any arbitrary point in continuum bodies:

$$\sigma_n = \sigma_{ij}n_in_j = \sigma_{11}n_1^2 + \sigma_{22}n_2^2 + \sigma_{33}n_3^2 = \sigma_1n_1^2 + \sigma_2n_2^2 + \sigma_3n_3^2 \tag{2.33}$$

Relation (2.32) indicates that all shear stresses are considered zero because the relation has been written only for the principal stress condition. According to relation (2.22), we know that:

$$\tau_n^2 = \left(T^{(n)}\right)^2 - \sigma_n^2$$

$$\left(T^{(n)}\right)^2 = T^{(n)}.T^{(n)} \equiv \sigma_{ij}\sigma_{ik}.n_j.n_k \equiv \sigma_1^2 n_1^2 + \sigma_2^2 n_2^2 + \sigma_3^2 n_3^2 \tag{2.34}$$

Again, note that the shear stresses are zero because we have considered only the principal planes.

$$\begin{aligned} \tau_n^2 &= \sigma_1^2 n_1^2 + \sigma_2^2 n_2^2 + \sigma_3^2 n_3^2 - (\sigma_1 n_1^2 + \sigma_2 n_2^2 + \sigma_3 n_3^2)^2 \\ &\equiv (\sigma_1 - \sigma_2)^2 n_1^2 n_2^2 + (\sigma_2 - \sigma_3)^2 n_2^2 n_3^2 + (\sigma_1 - \sigma_3)^2 n_1^2 n_3^2 \end{aligned} \tag{2.35}$$

Also, we recall the relation (1.14): $n_i.n_i = n_1^2 + n_2^2 + n_3^2 = 1$.

We aim to calculate the minimum and maximum values of shear stresses based on relation (2.35), but firstly, we should explain the Lagrange method in mathematics:

Assume a scalar function $F(x_1, x_2, x_3) = x_1^2 + x_2^2 + x_3^2$ and we want to find the extremums of this function under the condition $x_1 + x_2 + x_3 = 1$:

Let's rewrite F using the following equation:

$$F(x_1, x_2, x_3) = x_1^2 + x_2^2 + x_3^2 = x_1^2 + x_2^2 + x_3^2 + \lambda(x_1 + x_2 + x_3 - 1)$$

The Lagrange method expresses that the extremums of the function F are calculated by: $\frac{\partial F}{\partial x_1} = 0, \frac{\partial F}{\partial x_2} = 0, \frac{\partial F}{\partial x_3} = 0$

Thus:

$$\frac{\partial F}{\partial x_1} = 2x_1 + \lambda = 0, \ \frac{\partial F}{\partial x_2} = 2x_2 + \lambda = 0, \ \frac{\partial F}{\partial x_3} = 2x_3 + \lambda = 0$$

$$\to \left\{ \begin{array}{l} 2x_1 + \lambda = 0 \to x_1 = \frac{-\lambda}{2} \\ 2x_2 + \lambda = 0 \to x_2 = \frac{-\lambda}{2} \\ 2x_3 + \lambda = 0 \to x_3 = \frac{-\lambda}{2} \end{array} \right\} \to \begin{array}{c} x_1 + x_2 + x_3 = 1 \\ \frac{-\lambda}{2} - \frac{-\lambda}{2} - \frac{-\lambda}{2} = 1 \\ \lambda = \frac{-2}{3} \end{array}$$

Therefore, the maximum values of the function F are:

$$\lambda = -\frac{2}{3} \to \begin{array}{c} x_1 = \frac{1}{3} \\ x_2 = \frac{1}{3} \\ x_3 = \frac{1}{3} \end{array} \to x_1 = x_2 = x_3 = \frac{1}{3}$$

Note that $x_1 + x_2 + x_3 = \frac{1}{3} + \frac{1}{3} + \frac{1}{3} = 1$.

Now, we use the Lagrange method to find the extremums of the shear stress function τ_n:

Where $g(n_1, n_2, n_3) = n_1^2 + n_2^2 + n_3^2 = 1$

$$\begin{aligned} F(n_1, n_2, n_3, \lambda) &= \tau_n^2 + \lambda(g(n_1, n_2, n_3) - 1) \\ F(n_1, n_2, n_3, \lambda) &= \tau_n^2 + \lambda(n_1^2 + n_2^2 + n_3^2 - 1) \\ &\equiv \sigma_1^2 n_1^2 + \sigma_2^2 n_2^2 + \sigma_3^2 n_3^2 - (\sigma_1 n_1^2 + \sigma_2 n_2^2 + \sigma_3 n_3^2)^2 \\ &+ \lambda\left(n_1^2 + n_2^2 + n_3^2 - 1\right) \end{aligned}$$

$$\begin{cases} \frac{\partial F}{\partial n_1} = n_1\sigma_1^2 - 2n_1\sigma_1\left(\sigma_1 n_1^2 + \sigma_2 n_2^2 + \sigma_3 n_3^2\right) + \lambda n_1 = 0(A) \\ \frac{\partial F}{\partial n_2} = n_2\sigma_2^2 - 2n_2\sigma_2\left(\sigma_1 n_1^2 + \sigma_2 n_2^2 + \sigma_3 n_3^2\right) + \lambda n_2 = 0(B) \\ \frac{\partial F}{\partial n_3} = n_3\sigma_3^2 - 2n_3\sigma_3\left(\sigma_1 n_1^2 + \sigma_2 n_2^2 + \sigma_3 n_3^2\right) + \lambda n_3 = 0(C) \end{cases}$$

Now, n_1 is multiplied by equation A, n_2 is multiplied by equation B, and n_3 is multiplied by C. Then, considering the equation $\sigma_n = \sigma_1 n_1^2 + \sigma_2 n_2^2 + \sigma_3 n_3^2$, we write:

$$\begin{cases} n_1^2\sigma_1^2 - 2\sigma_1 n_1^2\sigma_n + n_1^2\lambda = 0(1) \\ n_2^2\sigma_2^2 - 2\sigma_1 n_2^2\sigma_n + n_2^2\lambda = 0(2) \\ n_3^2\sigma_3^2 - 2\sigma_1 n_3^2\sigma_n + n_3^2\lambda = 0(3) \end{cases}$$

$$(1) + (2) + (3) \rightarrow \left[n_1^2\sigma_1^2 + n_2^2\sigma_2^2 + n_3^2\sigma_3^2\right]$$
$$-2\left(\sigma_1 n_1^2 + \sigma_2 n_2^2 + \sigma_3 n_3^2\right)\sigma_n + \lambda\left(n_1^2 + n_2^2 + n_3^2\right) = 0$$

The above relation can be rewritten using relation (2.35):

$$\rightarrow \left[\tau_n^2 + \left(\sigma_1 n_1^2 + \sigma_2 n_2^2 + \sigma_3 n_3^2\right)^2\right] - 2\sigma_n^2 + \lambda = 0$$
$$\rightarrow \left[\tau_n^2 + \sigma_n^2\right] - 2\sigma_n^2 + \lambda = 0 \rightarrow \lambda = \sigma_n^2 - \tau n^2(4)$$

We substitute Eq. 4 in A, B, and C relations:

$$\begin{cases} \dfrac{\partial F}{\partial n_1} = n_1\sigma_1^2 - 2n_1\sigma_1\left(\sigma_1 n_1^2 + \sigma_2 n_2^2 + \sigma_3 n_3^2\right) + \left(\sigma_n^2 - \tau_n^2\right)n_1 = 0 \\ \equiv \left(\sigma_1^2 - 2\sigma_1\sigma_n + \sigma_n^2 - \tau_n^2\right)n_1 = 0 \\ \dfrac{\partial F}{\partial n_2} = \left(\sigma_2^2 - 2\sigma_2\sigma_n + \sigma_n^2 - \tau_n^2\right)n_2 = 0 \\ \dfrac{\partial F}{\partial n_3} = \left(\sigma_3^2 - 2\sigma_3\sigma_n + \sigma_n^2 - \tau_n^2\right)n_3 = 0 \end{cases}$$

One answer to the above equations is $n_1 = n_2 = n_3 = 0$ that cannot satisfy the condition $n_1^2 + n_2^2 + n_3^2 = 1$.

If we consider $n_1 = n_2 = 0$ and $n_3 \neq 0$:

$$n_1 = n_2 = 0 \rightarrow 0^2 + 0^2 + n_3^2 = 1 \rightarrow n_3 = \pm 1 \rightarrow \tau_n = 0$$

Similarly:

$$\begin{cases} n_1 = n_3 = 0 \rightarrow n_2 = \pm 1 \rightarrow \tau_n = 0 \\ n_1 = n_2 = 0 \rightarrow n_3 = \pm 1 \rightarrow \tau_n = 0 \end{cases}$$

We have seen these answers before, which state that the shear stress value τ_n is zero. This exactly conforms to the concept of principal planes, where the normal stress is maximum, and the shear stress is zero. Therefore, the minimum value of shear stress is zero. In the third case, we have:

$$\mathrm{n}_1 \neq 0$$

$$\mathrm{n}_2 \neq 0 \rightarrow \begin{cases} \frac{\partial F}{\partial n_2} = \sigma_2^2 - 2\sigma_2\sigma_n + \sigma_n^2 - \tau_n^2 = 0 \\ \frac{\partial F}{\partial n_3} = \sigma_3^2 - 2\sigma_3\sigma_n + \sigma_n^2 - \tau_n^2 = 0 \end{cases}$$

$$\mathrm{n}_3 \neq 0$$

We subtract both relations and:

$$\rightarrow \sigma_2^2 - \sigma_3^2 - 2\sigma_2\sigma_n + 2\sigma_3\sigma_n = 0 \rightarrow \sigma_2^2 - \sigma_3^2 - 2\sigma_n(\sigma_2 - \sigma_3) = 0$$

$$\rightarrow \sigma_2 + \sigma_3 = 2\sigma_n(D)$$

$$n_1 = 0 \rightarrow \begin{cases} \sigma_n = \sigma_1 n_1^2 + \sigma_2 n_2^2 + \sigma_3 n_3^2 = \sigma_2 n_2^2 + \sigma_3 n_3^2 (E) \\ n_1^2 + n_2^2 + n_3^2 = 1 \rightarrow n_2^2 + n_3^2 = 1(F) \end{cases}$$

$$(D)(E), (F) \rightarrow \begin{cases} \sigma_2 + \sigma_3 = 2\sigma_n \\ \sigma_2 + \sigma_3 = 2(\sigma_2 n_2^2 + \sigma_3 n_3^2) \\ \sigma_2 + \sigma_3 = 2(\sigma_2 n_2^2 + \sigma_3 (1 - n_2^2)) = 0 \end{cases}$$

$$\rightarrow n_2 = \pm\frac{1}{\sqrt{2}}$$

$$\rightarrow n_2 = \pm\frac{1}{\sqrt{2}} \rightarrow n_3 = \sqrt{1 - n_2^2} = \pm\frac{1}{\sqrt{2}}$$

$$\rightarrow \tau_n = (\sigma_2 - \sigma_3)^2 n_2^2 n_3^2 = \frac{\sigma_2 - \sigma_3}{2}$$

We repeat the calculations for $\begin{cases} \mathrm{n}_2 = 0, \mathrm{n}_1 \neq 0, \mathrm{n}_3 \neq 0 \\ \mathrm{n}_3 = 0, \mathrm{n}_1 \neq 0, \mathrm{n}_2 \neq 0 \end{cases}$ and obtain the maximum values of τ_n :

$$\begin{cases} n_1 = 0, n_2 = n_3 = \pm\frac{1}{\sqrt{2}}, \tau_n = \pm\frac{\sigma_2 - \sigma_3}{2} \\ n_1 = \pm\frac{1}{\sqrt{2}}, n_2 = 0, n_3 = \pm\frac{1}{\sqrt{2}}, \tau_n = \pm\frac{\sigma_1 - \sigma_3}{2} \\ n_1 = \frac{\pm 1}{\sqrt{2}}, n_2 = \frac{\pm 1}{\sqrt{2}}, n_3 = 0, \tau_n = \pm\frac{\sigma_1 - \sigma_3}{2} \end{cases}$$

$$n_1 = 0, n_2 = n_3 = \pm\frac{1}{\sqrt{2}}, \tau_n = \pm\frac{\sigma_2 - \sigma_3}{2}$$

$$n_1 = \pm\frac{1}{\sqrt{2}}, n_2 = 0, n_3 = \pm\frac{1}{\sqrt{2}}, \tau_n = \pm\frac{\sigma_1 - \sigma_3}{2}$$

$$n_1 = \pm\frac{1}{\sqrt{2}}, n_3 = 0, n_2 = \pm\frac{1}{\sqrt{2}}, \tau_n = \pm\frac{\sigma_1 - \sigma_2}{2}$$

We conclude that if the principal stresses are assumed as $\sigma_1 \geq \sigma_2 \geq \sigma_3$, then the maximum.

Shear stress value τ_n is calculated by:

$$\tau_n(max) = \frac{1}{2}|\sigma_1 - \sigma_3| \tag{2.36}$$

Stress Deviator Tensor

The stress tensor σ_{ij} can be decomposed into two other tensors as below:

A spherical (or hydrostatic) stress tensor that is basically equal in all directions, that is, just like hydrostatic pressure. It is calculated by calculating the average of principal stress:

$$\sigma_M = \frac{\sigma_1 + \sigma_2 + \sigma_3}{3} \equiv \frac{1}{3}\sigma_{KK}$$

The deviatoric stress is what is left after the spherical stress is taken out. This tensor intends to twist a continuum body which is under stress and named as S_{ij}:

Therefore, any stress tensor can be written based on A and B tensors:

$$\sigma_{ij} = S_{ij} + \frac{\sigma_{KK}}{3}\delta_{ij} \tag{2.37}$$

Now, we write the matrix form of spherical and deviator stress tensors:

$$[\sigma_M] = \frac{\sigma_{KK}}{3} = \begin{bmatrix} \sigma_M & 0 & 0 \\ 0 & \sigma_M & 0 \\ 0 & 0 & \sigma_M \end{bmatrix}$$

$$= \begin{bmatrix} \frac{\sigma_1+\sigma_2+\sigma_3}{3} & 0 & 0 \\ 0 & \frac{\sigma_1+\sigma_2+\sigma_3}{3} & 0 \\ 0 & 0 & \frac{\sigma_1+\sigma_2+\sigma_3}{3} \end{bmatrix}$$

$$S_{ij} = \begin{bmatrix} s_{11} & s_{12} & s_{13} \\ s_{21} & s_{22} & s_{23} \\ s_{31} & s_{32} & s_{33} \end{bmatrix} \begin{bmatrix} s_{11} & s_{12} & s_{13} \\ s_{21} & s_{22} & s_{23} \\ s_{31} & s_{32} & s_{33} \end{bmatrix}$$

$$= \begin{bmatrix} \sigma_{11} & \sigma_{12} & \sigma_{13} \\ \sigma_{21} & \sigma_{22} & \sigma_{23} \\ \sigma_{31} & \sigma_{32} & \sigma_{33} \end{bmatrix} - \begin{bmatrix} \sigma_M & 0 & 0 \\ 0 & \sigma_M & 0 \\ 0 & 0 & \sigma_M \end{bmatrix} \tag{2.38}$$

$$\begin{bmatrix} s_{11} & s_{12} & s_{13} \\ s_{21} & s_{22} & s_{23} \\ s_{31} & s_{32} & s_{33} \end{bmatrix} = \begin{bmatrix} \sigma_{11} - \sigma_M & \sigma_{12} & \sigma_{13} \\ \sigma_{21} & \sigma_{22} - \sigma_M & \sigma_{23} \\ \sigma_{31} & \sigma_{32} & \sigma_{33} - \sigma_M \end{bmatrix} \tag{2.39}$$

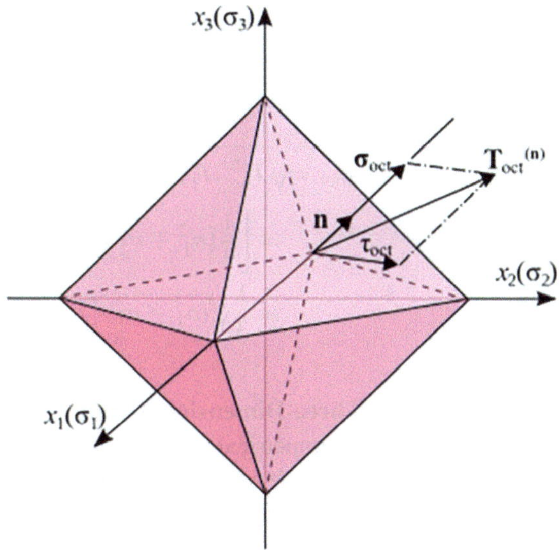

Fig. 2.8 The octahedral planes and related stress vectors

Octahedral Stresses

Figure 2.8 shows a octahedral plane. The characteristic of this planes is that their normal vector of each plane makes equal angles with the three principal stress directions. As can be seen in the figure, there are eight planes that have this property. The normal and shear components of the stress vector in these planes are called normal octahedral stress σ_{oct} and octahedral shear stress τ_{oct}.

Assume that the stress tensor at point O is written based on the principal directions according to Fig. 2.8:

$$\sigma_{ij} = \begin{bmatrix} \sigma_1 & 0 & 0 \\ 0 & \sigma_1 & 0 \\ 0 & 0 & \sigma_1 \end{bmatrix}$$

The stress vector in a octahedral plane $T_{oct}^{(n)}$ is calculated by:

$$\begin{matrix} n_1^2 + n_2^2 + n_3^2 = 1 \\ n_1 = n_2 = n_3 \end{matrix} \rightarrow n_1 = n_2 = n_3 = \frac{1}{\sqrt{3}}$$

$$\begin{aligned} T_{oct}^{(n)} =& \sigma_{ij}.n_i.e_j + \sigma_1.n_1.e_1 + \sigma_2.n_2.e_2.e_2 \\ & + \sigma_3.n_3.e_3 = \frac{1}{\sqrt{3}}(\sigma_1 e_1 + \sigma_2 e_2 + \sigma_3 e_3) \end{aligned} \tag{2.40}$$

The normal stress component of octahedral stress vector is defined by:

$$\sigma_{oct} = T_i^{(n)}.n_i \equiv \sigma_{ij}.n_i.e_j \equiv \sigma_1.n_1.n_1 + \sigma_2.n_2.n_2$$

$$+\sigma_3.n_3.n_3 = \frac{1}{\sqrt{3}}(\sigma_1+\sigma_2+\sigma_3) \tag{2.41}$$

The shear stress component of octahedral stress vector is calculated by:

$$\begin{aligned}\tau_{oct} &= \sqrt{T_i^{(n)}.T_i^{(n)} - \sigma_n^2} = \sqrt{T_{oct}^{(n)}.T_{oct}^{(n)} - \sigma_{oct}^2} \\ &= \left[\frac{1}{3}(\sigma_1^2+\sigma_2^2+\sigma_3^2) - \frac{1}{9}(\sigma_1+\sigma_2+\sigma_3)^2\right]^2 \\ &\equiv \frac{1}{3}\sqrt{(\sigma_1-\sigma_2)^2+(\sigma_2-\sigma_3)^2+(\sigma_3-\sigma_1)^2}\end{aligned} \tag{2.42}$$

Mohr's Circle in Three-Dimensional State

Mohr's circle is a geometrical solution to present the applied stress on a point which includes three circles. We recall relations (2.33), (2.34), and (2.35) once again which represent the principal stress tensor:

$$\begin{aligned}(T^{(n)})^2 &= \sigma_1^2 n_1^2 + \sigma_2^2 n_2^2 + \sigma_3^2 n_3^2 \\ \sigma_n &= \sigma_1 n_1^2 + \sigma_2 n_2^2 + \sigma_3 n_3^2 \\ \tau_n^2 &= (T^{(n)})^2 - \sigma_n^2 = \sigma_1^2 n_1^2 + \sigma_2^2 n_2^2 + \sigma_3^2 n_3^2 - \sigma_n^2\end{aligned}$$

In addition, we know that: $n_1^2 + n_2^2 + n_3^2 = 1$.

Now, we obtain n_1^2, n_2^2, n_3^2 based on the above equations:

$$\begin{aligned}n_1^2 &= \frac{\tau_n^2 + (\sigma-\sigma_2)(\sigma-\sigma_1)}{(\sigma_1-\sigma_2)(\sigma_1-\sigma_3)} \geq 0 \\ n_2^2 &= \frac{\tau_n^2 + (\sigma-\sigma_3)(\sigma-\sigma_1)}{(\sigma_2-\sigma_3)(\sigma_2-\sigma_1)} \geq 0 \\ n_3^2 &= \frac{\tau_n^2 + (\sigma-\sigma_1)(\sigma-\sigma_2)}{(\sigma_3-\sigma_1)(\sigma_3-\sigma_2)} \geq 0\end{aligned} \tag{2.43}$$

Now, if we consider $\sigma_1 \geq \sigma_2 \geq \sigma_3$ and noting that $n_1^2, n_2^2, n_3^2 \geq 0$ then we conclude that:

$$\sigma_1 \geq \sigma_2 \geq \sigma_3 \rightarrow (\sigma_1-\sigma_2)(\sigma_1-\sigma_3) > 0 \rightarrow \tau_n^2 + (\sigma-\sigma_2)(\sigma-\sigma_3) \geq 0$$

And also:

$$\begin{cases}\tau_n^2 + (\sigma-\sigma_3)(\sigma-\sigma_1) \leq 0 \\ \tau_n^2 + (\sigma-\sigma_1)(\sigma-\sigma_2) \geq 0\end{cases} \tag{2.44}$$

The above relations are rewritten as follows:

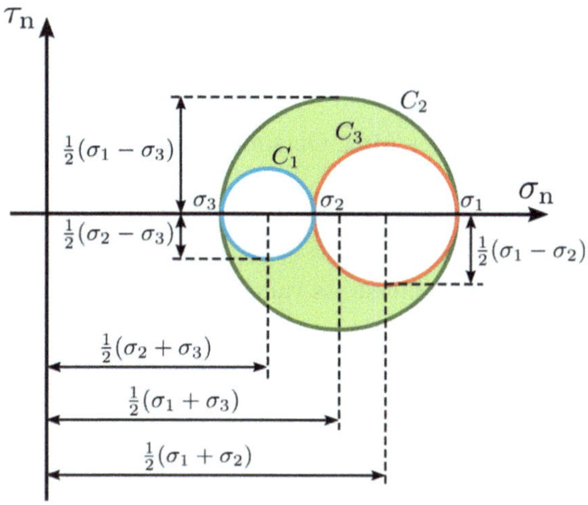

Fig. 2.9 The three-dimensional Mohr's circles

$$\begin{cases} \tau_n^2 + (\sigma - \sigma_2)(\sigma - \sigma_3) \geq 0 \rightarrow \tau_n^2 + \left[\sigma - \frac{1}{2}(\sigma_2 + \sigma_1)\right]^2 \geq \frac{1}{4}(\sigma_2 - \sigma_3)^2 \\ \tau_n^2 + (\sigma - \sigma_3)(\sigma - \sigma_1) \leq 0 \rightarrow \tau_n^2 + \left[\sigma - \frac{1}{2}(\sigma_1 + \sigma_3)\right]^2 \leq \frac{1}{4}(\sigma_3 - \sigma_1)^2 \\ \tau_n^2 + (\sigma - \sigma_1)(\sigma - \sigma_2) \geq 0 \rightarrow \tau_n^2 + \left[\sigma - \frac{1}{2}(\sigma_1 + \sigma_2)\right]^2 \geq \frac{1}{4}(\sigma_1 - \sigma_2)^2 \end{cases} \tag{2.45}$$

Or:

$$\begin{aligned} &\tau_n^2 + [\sigma - C_1]^2 \geq R_1^2 : \begin{cases} C_1 = \frac{1}{2}(\sigma_2 + \sigma_3) \\ R_1 = \frac{1}{2}(\sigma_2 - \sigma_3) \end{cases} \\ &\tau_n^2 + [\sigma - C_1]^2 \leq R_2^2 : \begin{cases} C_2 = \frac{1}{3}(\sigma_1 + \sigma_3) \\ R_2 = \frac{1}{2}(\sigma_3 - \sigma_1) \end{cases} \\ &\tau_n^2 + [\sigma - C_1]^2 \geq R_3^2 : \begin{cases} C_3 = \frac{1}{2}(\sigma_1 + \sigma_2) \\ R_3 = \frac{1}{2}(\sigma_1 - \sigma_2) \end{cases} \end{aligned} \tag{2.46}$$

Relation (2.46) represents three equations of circles with centers C_1, C_2, C_3, and radiuses of R_1, R_2, R_3. The circles are the Mohr's circles which are shown in Fig. 2.9. The circles indicate that if a stress point of (σ_n, τ_n) is satisfied by the circle equation of C_1 and R_1 when the point is located on or outside of the circle. The stress points that are satisfied by the circle equation of C_2 and R_2 should be located on or inside the circle. The stress points should be on or outside of circle C_3 and R_3 to be satisfied by the circle equation. In all circle equations, if the equation is equal to zero means that the stress points are on the circle and if the equation is grater/smaller than zero means that the stress points are outside/inside of the circle. We did not include the two dimensional Mohr's circle in this book as it could be read in all published strength of material books.

Sample Problems

1. The stress tensor of a point in the continuum environment is defined by

$\sigma_{ij} = \begin{bmatrix} 1\,4\,0 \\ 4\,1\,0 \\ 0\,0\,1 \end{bmatrix}$ Mpa. Obtain the following:

A-The principal stresses.

B-The maximum stress value.

C-The deviator tensor.

The octahedral stress vector, normal Octahedral stress, and shear octahedral stress.

A. To obtain the principal stress, we use $|\sigma_{ij} - \lambda\delta_{ij}| = 0$

$$\left|\begin{bmatrix} 1\,4\,0 \\ 4\,1\,0 \\ 0\,0\,1 \end{bmatrix} - \lambda\begin{bmatrix} 1\,0\,0 \\ 0\,1\,0 \\ 0\,0\,1 \end{bmatrix}\right| = \begin{vmatrix} 1-\lambda & 4 & 0 \\ 4 & 1-\lambda & 0 \\ 0 & 0 & 1-\lambda \end{vmatrix} = 0$$

$$\begin{vmatrix} 1-\lambda & 4 & 0 \\ 4 & 1-\lambda & 0 \\ 0 & 0 & 1-\lambda \end{vmatrix}\begin{bmatrix} 1-\lambda & 4 \\ 4 & 1-\lambda \\ 0 & 0 \end{bmatrix} = 0 \rightarrow^{Sarus} (1-\lambda)^3 - 16(1-\lambda) = 0$$

$$\rightarrow \begin{cases} \lambda_1 = 1Mpa \\ \lambda_2 = -3Mpa \\ \lambda_3 = 5Mpa \end{cases}$$

B. The principal stresses are ordered by $\sigma_1 > \sigma_2 > \sigma_3$:

$$5 > 1 > -3 \rightarrow \tau_{max} = \left|\frac{\sigma max - \sigma min}{2}\right| = \left|\frac{5-(-3)}{2}\right| = 4Mpa$$

C. The deviator tensor is obtained by $S_{ij} = \sigma_{ij} - \frac{\sigma_{kk}}{3}\delta_{ij}$

$$S_{ij} = \begin{bmatrix} 1\,4\,0 \\ 4\,1\,0 \\ 0\,0\,1 \end{bmatrix} - \begin{bmatrix} \sigma_M & 0 & 0 \\ 0 & \sigma_M & 0 \\ 0 & 0 & \sigma_M \end{bmatrix} = \begin{bmatrix} 1\,4\,0 \\ 4\,1\,0 \end{bmatrix} - \begin{bmatrix} 1\,0\,0 \\ 0\,1\,0 \\ 0\,0\,1 \end{bmatrix}$$

$$\sigma_M = \frac{\sigma_1 + \sigma_2 + \sigma_3}{3} = \frac{1-3+5}{3} = 1Mpa$$

$$S_{ij} = \begin{bmatrix} 0\,4\,0 \\ 4\,0\,0 \\ 0\,0\,0 \end{bmatrix}$$

D. Note that octahedral stress vector is calculated only based on principal directions therefore the stress tensor should be defined according to principal stress:

$$\sigma_{ij} = \begin{bmatrix} \lambda_1 & 0 & 0 \\ 0 & \lambda_1 & 0 \\ 0 & 0 & \lambda_1 \end{bmatrix} = \begin{bmatrix} 1 & 0 & 0 \\ 0 & -3 & 0 \\ 0 & 0 & 5 \end{bmatrix}$$

$$\begin{aligned} T_{oct}^{(n)} &= \sigma_{ij}.n_j.e_j = \frac{1}{\sqrt{3}}(\sigma_1 e_1 + \sigma_2 e_2 + \sigma_3 e_3) \\ &\equiv \frac{1}{\sqrt{3}} e_1 - \frac{3}{\sqrt{3}} e_2 + \frac{5}{\sqrt{3}} e_1 \end{aligned}$$

$$\sigma_{oct} \equiv T_i^{(n)}.n_i = \frac{1}{3}(\sigma_1 + \sigma_2 + \sigma_3) = \frac{1 - 3 + 5}{3} = 1Mpa$$

$$\begin{aligned} \tau_{oct} =& \frac{1}{3}\sqrt{(\sigma_1 - \sigma_2)^2 + (\sigma_2 - \sigma_3)^2 + (\sigma_3 - \sigma_1)^2} \\ & \frac{1}{3}\sqrt{(1 + 3)^2 + (-3 - 5)^2 + (5 - 1)^2} \\ =& \frac{\sqrt{16 + 64 + 16}}{3} = \frac{\sqrt{96}}{3} = \sqrt{\frac{32}{3}} Mpa \end{aligned}$$

2. Determine the principal direction for the biggest principal stress of the tensor

$$\sigma_{ij} = \begin{bmatrix} 2 & 1 & 1 \\ 1 & 2 & 1 \\ 1 & 1 & 2 \end{bmatrix} Mpa$$

$$\left|\sigma_{ij} - \lambda\delta_{ij}\right| = 0 \rightarrow \begin{bmatrix} 2-\lambda & 1 & 1 \\ 1 & 2-\lambda & 1 \\ 1 & 1 & 2-\lambda \end{bmatrix} = 0$$

$$\rightarrow \lambda^3 - 6\lambda^2 + 9\lambda - 4 = 0 \rightarrow \begin{cases} \lambda_1 = \sigma_1 = 1Mpa \\ \lambda_2 = \sigma_2 = 1Mpa \\ \lambda_3 = \sigma_3 = 4Mpa \end{cases}$$

The biggest principal stress is 4, and the related principal direction is calculated by: $(\sigma_{ij} - \lambda\delta_{ij})n_j = 0$

$$\begin{aligned} (\sigma_{11} - \lambda)n_1 + \sigma_{12}n_2 + \sigma_{13}n_3 = 0 \\ \sigma_{21}n_1 + (\sigma_{22} - \lambda)n_2 + (\sigma_{23})n_3 = 0 \\ \sigma_{31}n_1 + \sigma_{32}n_2 + (\sigma_{33} - \lambda)n_3 = 0 \end{aligned}$$

$$\lambda = 4 \rightarrow \begin{cases} (2-4)n_1 + n_2 + n_3 = 0 \\ n_1 + (2-4)n_2 + n_3 = 0 \\ n_1 + n_2 + (2-4)n_3 = 0 \end{cases}$$

$$\rightarrow \begin{cases} -2n_1 + n_2 + n_3 = 0 \\ n_1 - 2n_2 + n_3 = 0 \\ n_1 + n_2 - 2n_3 = 0 \end{cases}$$

$$\rightarrow \begin{cases} n_1 = n_3 \\ n_1 = n_2 \rightarrow \quad n_1 = n_2 = n_3 = \dfrac{\pm 1}{\sqrt{3}} \\ n_1^2 + n_2^2 + n_3^2 = 1 \end{cases}$$

$$\rightarrow \begin{cases} \vec{n} = \frac{i_1 + i_2 + i_3}{\sqrt{3}} \\ \vec{n} = \frac{-i_1 - i_2 - i_3}{\sqrt{3}} \end{cases}$$

3. The stress tensor of a point is given below. One of the principal stresses is 3, and there is a relation between the others as $\sigma_1 = 2\sigma_2$. Determine the other principal stresses and the magnitude of b.

$$\sigma_{ij} = \begin{bmatrix} 4\,b\,b \\ b\,7\,2 \\ b\,2\,4 \end{bmatrix} Mpa$$

$$I_1 = \sigma_{11} + \sigma_{22} + \sigma_{33} = \sigma_1 + \sigma_2 + \sigma_3 \rightarrow \sigma_{11} + \sigma_{22} + \sigma_{33} = 4 + 7 + 4 = 15Mpa$$

$$\sigma_3 = 3Mpa \rightarrow \sigma_1 + \sigma_2 = 15 - 3 = 12Mpa$$

$$\begin{cases} \sigma_1 + \sigma_2 = 12Mpa \\ \sigma_1 = 2\sigma_2 \end{cases} \rightarrow \begin{cases} \sigma_2 = 4Mpa \\ \sigma_1 = 8Mpa \end{cases}$$

$$|\sigma_{ij} - \lambda\delta_{ij}| = 0 \rightarrow \begin{bmatrix} 4-\lambda & b & b \\ b & 4-\lambda & 2 \\ b & 2 & 4-\lambda \end{bmatrix} = 0$$

$$\rightarrow (4-\lambda^2)(7-\lambda) + 2b^2 + 2b^2 - b^2(4-\lambda) - 4(4-\lambda) - b^2(7-\lambda) = 0$$

$$\lambda_3 = \sigma_3 = 3Mpa \rightarrow (4-3)^2(7-3) + 4b^2 - b^2(4-3) - 4(4-3) - b^2(7-3) = 0$$

$$b^2 = 0 \rightarrow b = 0$$

4. For the stress tensor given, which is applied to a point, there is a passing plane through the point, with a normal vector n, in which there is no stress in that plane. Find σ_{11} and the normal vector of the zero-stress plane.

$$\sigma_{ij} = \begin{bmatrix} \sigma_{11} & 2 & 1 \\ 2 & 0 & 2 \\ 1 & 2 & 0 \end{bmatrix} Mpa$$

$$T_j^{(n)} = \sigma_{ij}.n_i \text{or}\{ \begin{matrix} T_i^{(n)} = \sigma_{ji}.n_j \equiv \sigma_{ij}.n_j(A) \\ \sigma_{ij} = \sigma_{ji} \end{matrix}$$

$$\rightarrow \begin{bmatrix} T_1^{(n)} \\ T_2^{(n)} \\ T_3^{(n)} \end{bmatrix} = \begin{bmatrix} \sigma_{11} & 2 & 1 \\ 2 & 0 & 2 \\ 1 & 2 & 0 \end{bmatrix} \begin{bmatrix} n_1 \\ n_2 \\ n_3 \end{bmatrix} \equiv \begin{bmatrix} \sigma_{11}n_1 + 2n_2 + n_3 \\ 2n_1 + 2n_3 \\ n_1 + 2n_2 \end{bmatrix} \rightarrow$$

To find a zero-stress state, $T^{(n)}$ should be set to zero.

$$\begin{cases} T_1^{(n)} = \sigma_{11}n_1 + 2n_2 + n_3 \equiv 0 \\ T_2^{(n)} = 2n_1 + 2n_3 \equiv 0 \\ T_2^{(n)} = n_1 + 2n_2 \equiv 0 \end{cases}$$

$$\rightarrow \begin{cases} n_1 = -n_3 \\ n_1 = -2n_2 \end{cases}$$

$$n_1^2 + n_2^2 + n_3^2 = 1 \rightarrow n_1^2 + (\frac{-n_1}{2})^2 + (-n_1)^2 = 1$$

$$n_1^2 + \frac{n_1^2}{4} + n_1^2 = 1$$

$$n_1^2\left(\frac{9}{4}\right) = 1 \rightarrow n_1^2 = \frac{4}{9} \rightarrow n_1 = \frac{\pm 2}{3}$$

$$\rightarrow \begin{cases} n_2 = \frac{\mp 1}{3} \\ n_3 = \frac{\mp 2}{3} \end{cases}$$

$$\rightarrow \vec{n} = \frac{2}{3}i_1 - \frac{1}{3}i_2 - \frac{2}{3}i_2$$

$$\rightarrow \sigma_{11}n_1 + 2n_2 + n_3 = 0 \rightarrow \sigma_{11}n_1 + 2\left(\frac{-1}{2}n_1\right) - n_1 = 0$$

$$\rightarrow \sigma_{11}n_1 - n_1 - n_1 = 0 \rightarrow \sigma_{11} \equiv \frac{2n_1}{n_1} = 2$$

$$\sigma_{11} = 2\,\text{Mpa}$$

5. For the given stress tensor below, find:

A. The stress vector applied to a point that is a member of a plane with normal vector

$$\vec{n} = \frac{1}{9}(\mathrm{i}_1 + 4\mathrm{i}_2 + 8\mathrm{i}_3)$$

B. Compute the size of the stress vector

C. What is the magnitude of the normal stress component

D. Calculate the angle between the stress vector and the normal vector

$$\sigma_{ij} = \begin{bmatrix} -9 & 3 & -6 \\ 3 & 6 & 9 \\ -6 & 9 & -6 \end{bmatrix}$$

$$T_i^{(n)} = \sigma_{ij}.n_j \rightarrow \begin{bmatrix} T_1 \\ T_2 \\ T_3 \end{bmatrix} = \begin{bmatrix} -9 & 3 & -6 \\ 3 & 6 & 9 \\ -6 & 9 & -6 \end{bmatrix} \begin{bmatrix} \frac{1}{9} \\ \frac{4}{9} \\ \frac{8}{9} \end{bmatrix}$$

(A)

$$\rightarrow \begin{cases} T_1 = \frac{-9}{9} + \frac{12}{9} - \frac{48}{9} = \frac{-45}{9} = -5Mpa \\ T_2 = \frac{-39}{9} + \frac{24}{9} + \frac{72}{9} = \frac{99}{9} = 11Mpa \\ T_3 = \frac{-6}{9} + \frac{36}{9} - \frac{48}{9} = \frac{-18}{9} = -2Mpa \end{cases}$$

$$\rightarrow T^{(n)} = -5i_1 + 11i_2 - 2i_3$$

(B) $\left|T^{(n)}\right| = \sqrt{(-5)^2 + (11)^2 + (-2)^2} = \sqrt{25 + 121 + 4} = \sqrt{150}Mpa$

(C)

$$\sigma_n = T^{(n)}.n \equiv (-5i_1 + 11i_2 - 2i_3).\left(\frac{i_1}{9} + \frac{4}{9}i_2 + \frac{8}{9}i_3\right)$$

$$\equiv \frac{-5}{9} + \frac{44}{9} - \frac{16}{9} \equiv \frac{23}{9}Mpa$$

(D)
$$T^{(n)}.n = \left|T^{(n)}\right|.|n|.Cos\theta \rightarrow \frac{23}{9} = \sqrt{150} \times 1 \times Cos\theta$$
$$\rightarrow Cos\theta = \frac{23/9}{\sqrt{150} \times 1} \rightarrow \theta = 77.94^{\circ}$$

6. For the given stress tensor:
   A. Show that the Cauchy equilibrium equations are satisfied, assuming that there is no volumetric force.
   B. Find the stress vector that is applied on the point P(2, −1, 6) from the plane

$$3x_1 + 6x_2 + 2x_3 = 12$$

$$\sigma_{ij} = \begin{bmatrix} (1 - x_1^2)x_2 + \frac{2}{3}x_2^3 & -(4 - x_2^2)x_1 & 0 \\ -(4 - x_2^2)x_1 & \frac{-1}{3}(x_2^3 - 12x_2) & 0 \\ 0 & 0 & (3 - x_1^2)x_2 \end{bmatrix}$$

Using relation (2.27):

$$\frac{\partial\sigma_{11}}{\partial x_1} + \frac{\partial\sigma_{12}}{\partial x_2} + \frac{\partial\sigma_{13}}{\partial x_3} + F_1 = 0 \rightarrow^{F_1=0}$$

$$\frac{\partial}{\partial x_1}\left[(1 - x_1^2)x_2 + \frac{2}{3}x_2^3\right] + \frac{\partial}{\partial x_2}\left[-(4 - x_2^2)x_1\right]$$

$$+\frac{\partial}{\partial x_3}[0] = -2x_1x_2 + 2x_2x_1 = 0 \rightarrow 0 = 0$$

$$\frac{\partial\sigma_{21}}{\partial x_1} + \frac{\partial\sigma_{22}}{\partial x_2} + \frac{\partial\sigma_{23}}{\partial x_3} + F_2 = 0 \rightarrow^{F_2=0}$$

$$\frac{\partial}{\partial x_1}\left[-(4 - x_2^2)x_1\right] + \frac{\partial}{\partial x_2}\left[\frac{-1}{3}(x_2^3 - 12x_2)\right] + \frac{\partial}{\partial x_3}(0) = 0$$

$$\rightarrow -4 + x_2^2 - x_2^2 + 4 = 0 \rightarrow 0 = 0$$

$$\frac{\partial\sigma_{31}}{\partial x_1} + \frac{\partial\sigma_{32}}{\partial x_2} + \frac{\partial\sigma_{33}}{\partial x_3} + F_3 = 0 \rightarrow^{F_3=0}$$

$$\frac{\partial}{\partial x_1}[0] + \frac{\partial}{\partial x_1}[0] + \frac{\partial}{\partial x_1}\left[(3 - x_1^2)x_2\right] = 0$$

$$\rightarrow 0 + 0 + 0 = 0 \rightarrow 0 = 0$$

Therefore, the Cauchy equilibrium equations are satisfied.

B. The plane has a unit normal vector $\overrightarrow{n} = \frac{(3i_1+6i_2+2i_3)}{7}$, but we first calculate the stress tensor components as per point P values:

$$\sigma_{ij} = \begin{bmatrix} (1-2^2)(-1)+\frac{2}{3}(-1)^3 & -(4-(-1)^2)2 & 0 \\ -(4-(-1)^2)(2) & \frac{-1}{3}((-1)^3-12(-1)) & 0 \\ 0 & 0 & (3-(2)^2)(-1) \end{bmatrix}$$

$$= \begin{bmatrix} \frac{7}{3} & -6 & 0 \\ -6 & -\frac{11}{3} & 0 \\ 0 & 0 & 1 \end{bmatrix}$$

$$T_j^{(n)} = \sigma_{ij}.n_j \rightarrow \begin{bmatrix} T_1 \\ T_2 \\ T_3 \end{bmatrix} = \begin{bmatrix} \frac{7}{3} & -6 & 0 \\ -6 & -\frac{11}{3} & 0 \\ 0 & 0 & 1 \end{bmatrix} \begin{bmatrix} \frac{3}{7} \\ \frac{6}{7} \\ \frac{2}{7} \end{bmatrix}$$

$$\vec{n} = \frac{(3i_1+6i_2+2i_3)}{7}$$

$$= \frac{1}{7}\begin{bmatrix} \frac{21}{3} - 36 \\ -18 - \frac{66}{3} \\ 2 \end{bmatrix}$$

$$\equiv \frac{1}{7}(-29i_1 - 40i_2 + 2i_3)$$

$$\rightarrow T_i^{(n)} \equiv \frac{1}{7}(-29i_1 - 40i_2 + 2i_3)$$

There is an important point: In the main Cauchy stress relation $T_i^{(n)} = \sigma_{ij}.n_j$, n_j is the normal unit vector, which is calculated by the following method:

$$ax + by + cz + d = 0 \rightarrow \vec{n} = (a, b, c)$$

$$\vec{n}_j = \frac{a_i + b_j + c_k}{\sqrt{a^2+b^2+c^2}}$$

Therefore, please always use the unit vector in the calculations.

7. The stress tensor for point P is given below, where a and b are unknown. If the principal stress tensor for the point P is $\sigma_{ij}^* = \begin{bmatrix} \sigma_1 & 0 & 0 \\ 0 & 2 & 0 \\ 0 & 0 & \sigma_3 \end{bmatrix}$ and the maximum shear stress value is 5.5 MPa, then find $\sigma_1 and \sigma_3$

$$\sigma_{ij} = \begin{bmatrix} 5 & a & -a \\ a & 0 & b \\ -a & b & 0 \end{bmatrix}$$

$$I_1 = \sigma_1 + \sigma_2 + \sigma_3 = \sigma_{11} + \sigma_{22} + \sigma_{33}$$

$$\rightarrow \sigma_1 + \sigma_2 + \sigma_3 = 5 + 0 + 0 \rightarrow \sigma_1 + \sigma_3 = 3Mpa(1)$$

$$\tau_{max} = \frac{\sigma_1 - \sigma_3}{2} \rightarrow \sigma_1 - \sigma_3 = 2\tau max = 2(5.5) = 1(2)$$

$$1, 2 \rightarrow \sigma_1 = 7Mpa, \sigma_3 = -4Mpa$$

8. For the given stress tensor, decompose it into spherical and deviator tensors, and find the principal stress values of the deviator tensor.

$$\sigma_{ij} = \begin{bmatrix} 29 & 0 & 0 \\ 0 & -26 & 6 \\ 0 & 6 & 9 \end{bmatrix}^{Mpa}$$

$$\begin{cases} \sigma_M = \frac{\sigma_{11}+\sigma_{22}+\sigma_{33}}{3} \equiv \frac{\sigma_1+\sigma_2+\sigma_3}{3} \rightarrow \sigma_M = \frac{29-26+9}{3} = 4Mpa \\ I_1 = \sigma_1 + \sigma_1 + \sigma_1 = \sigma_{11} + \sigma_{22} + \sigma_{33} \end{cases}$$

$$\rightarrow \sigma_M = \begin{bmatrix} 4 & 0 & 0 \\ 0 & 4 & 0 \\ 0 & 0 & 4 \end{bmatrix} \equiv \frac{\sigma_{KK}}{3}\delta_{ij}$$

$$S_{ij} = \sigma_{ij} - \frac{\sigma_{KK}}{3}\delta_{ij} \equiv \begin{bmatrix} 29 & 0 & 0 \\ 0 & -26 & 6 \\ 0 & 6 & 9 \end{bmatrix} - \begin{bmatrix} 4 & 0 & 0 \\ 0 & 4 & 0 \\ 0 & 0 & 4 \end{bmatrix} = \begin{bmatrix} 25 & 0 & 0 \\ 0 & -30 & 6 \\ 0 & 6 & 5 \end{bmatrix}$$

$$|\sigma_{ij} - \lambda\delta_{ij}| = 0 \rightarrow \begin{bmatrix} 25-\lambda & 1 & 1 \\ 0 & -30-\lambda & 6 \\ 0 & 6 & 5-\lambda \end{bmatrix} = 0$$

$$\begin{matrix} \rightarrow (25-\lambda)[(-30-\lambda)(5-\lambda) - 36] = 0 \rightarrow \\ (25-\lambda)\left[\left(\lambda^2 + 25\lambda - 186\right)\right] = 0 \rightarrow \end{matrix} \begin{cases} \lambda_1 = 25Mpa \\ \lambda_2 = 6Mpa \\ \lambda_3 = -31Mpa \end{cases}$$

9. The stress tensor is given below for a point in a continuum environment.

$$\sigma_{ij} = \begin{bmatrix} 0 & 0 & Ax_2 \\ 0 & 0 & -Bx_3 \\ Ax_2 & -Bx_3 & 0 \end{bmatrix}$$

A. If the system is under equilibrium, find the volumetric forces vector.
B. Principal stress, maximum shear stress, and stress invariants at point (0, B, A)

A.

$$\frac{\partial \sigma_{11}}{\partial x_1} + \frac{\partial \sigma_{12}}{\partial x_2} + \frac{\partial \sigma_{13}}{\partial x_3} + F_1 = 0 \rightarrow 0 + 0 + 0 + F_1 = 0 \rightarrow F_1 = 0$$

$$\frac{\partial \sigma_{21}}{\partial x_1} + \frac{\partial \sigma_{22}}{\partial x_2} + \frac{\partial \sigma_{23}}{\partial x_3} + F_2 = 0 \rightarrow 0 + 0 - B + F_2 = 0 \rightarrow F_2 = B$$

$$\frac{\partial \sigma_{31}}{\partial x_1} + \frac{\partial \sigma_{32}}{\partial x_2} + \frac{\partial \sigma_{33}}{\partial x_3} + F_3 = 0 \rightarrow 0 + 0 + 0 + F_3 = 0 \rightarrow F_3 = 0$$

$$\rightarrow \vec{B} = Be_2$$

$$0, B, A) \rightarrow \begin{cases} x_2 = B \\ x_3 = A \end{cases} \rightarrow \sigma_{ij} = \left[\begin{bmatrix} 0 & 0 & AB \\ 0 & 0 & -BA \\ AB & -BA & 0 \end{bmatrix}\right]$$

B.

$$|\sigma_{ij} - \lambda\delta_{ij}| = 0 \rightarrow \begin{bmatrix} -\lambda & 1 & AB \\ 0 & --\lambda & -BA \\ AB & -BA & -\lambda \end{bmatrix} = 0$$

$$\rightarrow -\lambda^3 + (BA)^2\lambda + (AB)^2\lambda = 0$$

$$\rightarrow \begin{cases} \lambda_1 = 0 \\ \lambda_1 = \sqrt{2}BA \\ \lambda_1 = -\sqrt{2}BA \end{cases}$$

C.

$$\begin{cases} I_1 = \sigma_1 + \sigma_2 + \sigma_3 = 0 \\ I_2 = \sigma_1\sigma_2 + \sigma_2\sigma_3 + \sigma_3\sigma_1 = 0 + \left(\sqrt{2}BA\right)\left(-\sqrt{2}BA\right) + 0 \equiv 2B^2A^2 \\ I_1 = \sigma_1\sigma_2\sigma_3 = 0 \end{cases}$$

10. The stress tensor of a point in a continuum body is given below. Assume there is a plane where the magnitude of normal stress is 12 MPa and the shear stress magnitude is 3 MPa. Determine the normal vector of the plane.

$$[\sigma_{ij}] = \begin{bmatrix} 9 & 3 & 0 \\ 3 & 9 & 0 \\ 0 & 0 & 18 \end{bmatrix}$$

First, we calculate the principal stresses:

$$\left|\sigma_{ij} - \lambda\delta_{ij}\right| = 0 \rightarrow \begin{bmatrix} 9-\lambda & 1 & 0 \\ 3 & 9-\lambda & 0 \\ 0 & 0 & 18-\lambda \end{bmatrix} = 0$$

$$\rightarrow (9-\lambda)^2(18-\lambda) - 9(18-\lambda) = (18-\lambda)(9-\lambda+3) =$$

$$\rightarrow \lambda_1 = 18 = \sigma_1$$

$$\rightarrow \lambda_2 = 6 = \sigma_2$$

$$\rightarrow \lambda_3 = 6 = \sigma_3$$

Now, we compute the magnitude of normal and shear stresses for principal stress tensor:

$$\begin{cases} \sigma_n = \sigma_1 n_1^2 + \sigma_2 n_1^2 + \sigma_3 n_1^2 \\ T^{(n)^2} = \tau_n^2 + \sigma_n^2 = \sigma_1^2 n_1^2 + \sigma_2^2 n_2^2 + \sigma_3^2 n_3^2 \end{cases}$$

$$\rightarrow \begin{cases} \sigma_N = 12^{Mpa} = \sigma_1 n_1^2 + \sigma_2 n_2^2 + \sigma_3 n_3^2 \equiv 18n_1^2 + 6n_2^2 + 12n_3^2 \\ \tau_n^2 + \sigma_n^2 = 3^2 + 12^2 \equiv 18^2 n_1^2 + 6^2 n_2^2 + 12^2 n_3^2 \end{cases}$$

$$\rightarrow \begin{cases} 18n_1^2 + 6n_2^2 + 12n_3^2 = 12 \\ 324n_1^2 + 36n_2^2 + 144n_3^2 = 153 \\ n_1^2 + n_2^2 + n_3^2 = 1 \end{cases} \rightarrow \begin{cases} n_1 = \frac{\pm 1}{\sqrt{8}} \\ n_2 = \pm\sqrt{\frac{3}{4}} \\ n_3 = \pm\sqrt{\frac{1}{8}} \end{cases}$$

$$\vec{n} = \sqrt{\frac{1}{8}} i_1 + \sqrt{\frac{3}{4}} i_2 + \sqrt{\frac{1}{8}} i_3$$

Theory of Deformation 3

When a continuum environment is affected by an external force or any other factor such as temperature or pressure variations then the environment is deformed. If the external factor varies with the time, then the deformation of the continuum environment is also time dependent. In this chapter, we intend to study the motion and deformation of a continuum environment without noting the driving factors for those deformations. Therefore, we focus on the deformation of the environment's particles and then investigate the total deformation of the continuum environment. We have tried to use relatively simplified equations to let the readers understand the theory of deformation easily.

The Deformation of a Continuum Environment

Figure 3.1 presents a continuum environment R0 that has been under effect of an external factor from $t = t_0$ and is deformed. Then it moves to another position and is changed to a new environment R. To study the geometries of the environments R0 and R, we use the Cartesian coordinates systems (X_1, X_2, X_3), (x_1, x_2, x_3) and assume they coincide (with a common coordinate origin point O). Assume point P which belongs to R0, moves to point p belonging to R during $t = 0$ to $t = t0$. It means that we want to study the movement of a point from the original environment to the new one and define it by vectors $\overrightarrow{OP}$ at $t = 0$ and $\overrightarrow{Op}$ at $t = t_0$.

If we present $\overrightarrow{OP}$ as X, $\overrightarrow{Op}$ as x, and u as the displacement vector between P and p then:

$$\vec{O}\mathrm{P} + \vec{\mathrm{u}} = \overrightarrow{\mathrm{Op}}\ \text{يا}\ X += x \tag{3.1}$$

where $x = x_1 i_1 + x_2 i_2 + x_3 i_3$ and $X = X_1 i_1 + X_2 i_2 + X_3 i_3$

K. Hassani and F. Azimifar, *Introduction to Continuum Mechanics*, Synthesis Lectures on Mechanical Engineering, https://doi.org/10.1007/978-3-032-05076-2_3

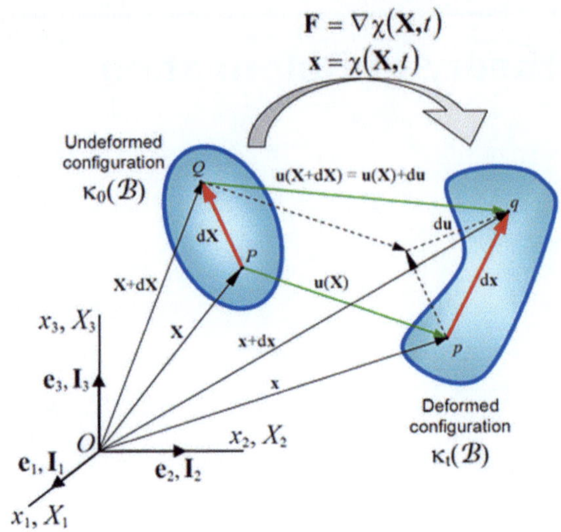

Fig. 3.1 The motion of the continuum environment during the time and its deformation

The whole process of motion is time dependent, therefore:

$$
\begin{aligned}
x &= x(X, t_\circ) \\
x &= x(X, t)
\end{aligned}
\tag{3.2}
$$

In the time t, we have:

$$
x = x(X, t)\text{or}\begin{cases} x_1 = x_1(X_1, X_2, X_3, t) \\ x_2 = x_2(X_1, X_2, X_3, t) \\ x_3 = x_3(X_1, X_2, X_3, t) \end{cases}
\tag{3.3}
$$

The relations (3.1) to (3.3) introduce a coordination system namely Lagrangian which expresses that the motion of each particle during the time is fully dependent on its origin position. Thus, when we want to study the motion of a particle during the time, we can use its origin position and the time to do that. For example, relation (3.2) shows that the particle is in X position, as starting point, at t = t0 and its motion can be tracked in any time t using Lagrangian coordinate based on X and t.

There are many functions that can describe the usage of Lagrangian coordinates as below:

$$
\begin{cases} x_1 = X_1 + t^2 X_2 \\ x_2 = X_2 + t^2 X_1 \\ x_3 = X_3 \end{cases}
\tag{3.4}
$$

Relation (3.4) exactly expresses where the position of a particle is, with new coordinates (x_1, x_2, x_3) at the time t. It is enough to have the initial position of the motion,

(X_1, X_2, X_3), and then we can obtain the new position which depends on the initial coordinates, X, and the time t.

Assume there is a particle in initial position (1, 2, 3) at t = 0 and we want to find its position after deformation of the continuum environment at t = 1 s

$$(X_1, X_2, X_3)_{t=1} \to (1, 2, 3) \to \begin{cases} x_1 = 1 + 1^2(2) = 3 \\ x_2 = 2 + 1^2(1) = 3 \\ x_3 = 3 \end{cases} \tag{3.5}$$

Therefore, the particle is in (3, 3, 3) at t = 1 s.

Lagrangian coordinates are used mainly in solid mechanics where we intend to investigate the motion path, velocity, and acceleration of a particle during deformation and motion of a continuum environment.

$$\begin{aligned} V(x,t) &= \frac{\partial x}{\partial t} = \overset{\circ}{x} \equiv V(X_1, X_2, X_3) \\ a(x,t) &= \frac{\partial^2 x}{\partial t} = \frac{\partial V}{\partial t} = \overset{\circ}{V} = \overset{\circ\circ}{x} = a(X_1, X_2, X_3, t) \end{aligned} \tag{3.6}$$

Relation (3.6) shows how to calculate the velocity and acceleration of a particle in Lagrangian coordinates. See the following which is continue of relation (3.5) example:

$$\begin{cases} x_1 = X_1 + t^2 X_2 \\ x_2 = X_2 + t^2 X_1 \\ x_3 = X_3 \end{cases} \to \begin{cases} \overset{\circ}{x}_1 = V_1 = 2tX_2 \\ \overset{\circ}{x}_2 = V_2 = 2tX_1 \\ \overset{\circ}{x}_3 = V_3 = 0 \end{cases} \tag{3.7}$$

The derivation has been done relating to time where (X_1, X_2, X_3) are independent of time and therefore their derivative relating to time is zero.

$$\begin{aligned} \overset{\circ\circ}{x_1} &= a_1 = 2X_2 \\ \overset{\circ\circ}{x_2} &= a_2 = 2X_1 \\ \overset{\circ\circ}{x_3} &= a_3 = 0 \end{aligned} \tag{3.8}$$

As can be seen, the velocity and acceleration of a particle can be obtained by finding the first and second derivation of the motion functions relating to time (Lagrangian coordinates). In fluid mechanics, it is more important to determine the velocity of passing particles from a special points and tracking of one particle is not useful. Therefore, we need a new coordinate system namely spatial or Eulerian coordinates. In Lagrangian coordinates, we defined a function, $x = K(X, t)$, which could describe the different positions of a particle depending on their initial position, X, and the time. The function is continuous and invertible:

$$x = k(X, t) \to X = K^{-1}(x, t) \tag{3.9}$$

Relation (3.9) is the mathematical expression for Eulerian coordinates which specifies the particle's specifications including velocity and acceleration in a special position, $x(x_1, x_2, x_3)$, at time t. The initial position of that particle is (X_1, X_2, X_3). See the following example:

$$\begin{cases} x_1 = X_1 + t^2 X_2 \\ x_2 = X_2 + t^2 X_1 \\ x_3 = X_3 \end{cases} \rightarrow \begin{cases} X_1 = \frac{x_2 t^2 - x_1}{t^4 - 1} \\ X_2 = \frac{x_1 t^2 - x_2}{t^4 - 1} \\ X_3 = x_3 \end{cases} \tag{3.10}$$

Relation (3.10) is the inversion of function $x = K(X, t)$ which is $X = K^{-1}(X, t)$. Now, we want to find the velocity and acceleration of a particle in a special position x_1, x_2, x_3 at time t. The initial position of the point was X_1, X_2, X_3 and we note that the Eulerian velocity and acceleration do not depend on X anymore and defined by $V(x, t), a(x, t)$. The derivative operator is defined by following equation (see reference 2 for more details)

$$\begin{cases} \frac{D}{Dt} = \frac{\partial}{\partial t} + \nabla . V = \frac{\partial}{\partial t} + V_i . \frac{\partial}{\partial x_i} \\ V(x, t) = \frac{\partial x}{\partial t} \end{cases} \tag{3.11}$$

It is useful to know that relation (3.11) was extracted using the chain derivative rule where Vi is the particle velocity in Lagrangian coordinates. Now, we calculate the Eulerian velocity and acceleration of the particle based on example of relation (3.10):

$$\begin{cases} X_1 = \frac{x_2 t^2 - x_1}{t^4 - 1} \\ X_2 = \frac{x_1 t^2 - x_2}{t^4 - 1} \\ X_3 = x_3 \end{cases} \rightarrow \begin{cases} \overset{\circ}{x_1} = V_1 = 2tX_2 = 2t\left(\frac{x_1 t^2 - x_2}{t^4 - 1}\right) \\ \overset{\circ}{x_2} = V_2 = 2tX_1 = 2t\left(\frac{x_2 t^2 - x_1}{t^4 - 1}\right) \\ \overset{\circ}{x_3} = V_3 = 0 \end{cases}$$

$$\begin{cases} \overset{\circ\circ}{x_1} = a_1 = \frac{2x_1 t^2 - 2x_2}{t^4 - 1} \\ \overset{\circ\circ}{x_2} = a_2 = \frac{2x_2 t^2 - 2x_1}{t^4 - 1} \\ \overset{\circ\circ}{x_3} = a_3 = 0 \end{cases} \tag{3.12}$$

Relation (3.12) express the velocity and acceleration of a particle at a special point like (x_1, x_2, x_3) at the time t. The relations say when a particle of fluid reaches a special position in space at the time t, then what velocity and acceleration are, and these parameters do not depend on the initial position of the particle at all, and this is the Eulerian coordinates concept. The concepts of both Lagrangian and Eulerian coordinates are very important, but we state below some supplementary explanations:

Lagrangian Concept

In this coordinate system, we can track point P every moment regarding its initial position $P_\circ$ and derive the velocity and acceleration of that point in that moment.

Eulerian Concept

In this coordinate, we can specify each particle's velocity and acceleration, P_0, P, P_2, when they reach a special position like (x_1, x_2, x_3) at time t.

Gradient of Deformation

Gradient of deformation is a tensor form which is derived by extracting the derivative of the vector elements $x = f(X, t)$ relative to X where x is the deformed vector, X is the initial reference vector and t is the time. Therefore, we define the gradient of deformation as:

$$F_{ij} = \frac{\partial x_i}{\partial X_j} = x_{i,j} = \begin{bmatrix} \frac{\partial x_1}{\partial X_1} & \frac{\partial x_1}{\partial X_2} & \frac{\partial x_1}{\partial X_3} \\ \frac{\partial x_2}{\partial X_1} & \frac{\partial x_2}{\partial X_2} & \frac{\partial x_2}{\partial X_3} \\ \frac{\partial x_3}{\partial X_1} & \frac{\partial x_3}{\partial X_2} & \frac{\partial x_3}{\partial X_3} \end{bmatrix} \tag{3.13}$$

We know $x = u + Xx = u + X$

$$\rightarrow F = \frac{\partial}{\partial X_j}(x_i) = \frac{\partial}{\partial X_j}(u_i + X_i) = \frac{\partial X_i}{\partial X_j} + \frac{\partial u_i}{\partial X_j} = \delta_{ij} + \frac{\partial u_i}{\partial X_j}$$

$$\text{or } F_{ij} = \delta_{ij} + u_{i,j} \tag{3.14}$$

Relation (3.13) is written in matrix form as below:

$$F_{ij} = \begin{bmatrix} 1 + \frac{\partial u_1}{\partial X_1} & \frac{\partial u_1}{\partial X_2} & \frac{\partial u_1}{\partial X_3} \\ \frac{\partial u_2}{\partial X_1} & 1 + \frac{\partial u_2}{\partial X_2} & \frac{\partial u_2}{\partial X_3} \\ \frac{\partial u_3}{\partial X_1} & \frac{\partial u_3}{\partial X_2} & 1 + \frac{\partial u_3}{\partial X_3} \end{bmatrix} \tag{3.15}$$

The gradient of deformation tensor is a linear transformation that transforms one element of the continuum body to a deformed element. Now, the differential form of the deformed vector, x, is written based on the derivative chain rule:

$$dx_k = \frac{\partial x_K}{\partial X_i}.dX_i = F_{K,i}.dX_i \text{ or } d\vec{x} = F.\overrightarrow{dX} \tag{3.16}$$

Relation (3.15) is rewritten in matrix form as below:

$$\begin{bmatrix} dx_1 \\ dx_2 \\ dx_3 \end{bmatrix} = \begin{bmatrix} \frac{\partial x_1}{\partial X_1} & \frac{\partial x_1}{\partial X_2} & \frac{\partial x_1}{\partial X_3} \\ \frac{\partial x_2}{\partial X_1} & \frac{\partial x_2}{\partial X_2} & \frac{\partial x_2}{\partial X_3} \\ \frac{\partial x_3}{\partial X_1} & \frac{\partial x_2}{\partial X_2} & \frac{\partial x_3}{\partial X_3} \end{bmatrix} \begin{bmatrix} dX_1 \\ dX_2 \\ dX_3 \end{bmatrix} \tag{3.17}$$

F_{ij} is sometimes called Jacoby matrix, J, and determinant of F is named Jacobien. If F = I, then the determinant of F is one and means that there is no deformation and the body is rigid completely.

Strain

Strain is one of the important concepts in deformation of the materials which presents the deformation of a continuum environment regarding the relative movement of its particles and there is no strain for a rigid environment. There are various definitions for strain in mechanical engineering, but engineering strain is more applicable one which is used for small strains. For bigger strains, bigger than 1%, other strain concepts are used such as stretch, Green, Almansi, and logarithm.

Cauchy strain or engineering strain is defined as the ratio of the total deformed length of the body, under stress, to the initial length. Assume, a continuum body is under an axial load then the engineering strain is calculated by the ratio of variation of the length of the body (ΔL) to its initial length (L):

$$e = \frac{\Delta L}{L_1} = \frac{L_2 - L_1}{L_1} \tag{3.18}$$

Strain is dimensionless and expressed as meter per meter or inch per inch. For example, if a rod with a length of 1 m has 0.0001 m of increase in its length under an axial load, the occurred strain is equal 0.0001. For some materials that exhibit bigger strains under the loading, Elastomers for instance, we use the extension ratio concept which is a standard for measuring the extension strain of a linear element and defined as the ratio of deformed length L2 to initial length L1:

$$\lambda = \frac{L_2}{L_1} \tag{3.19}$$

The relationship between the extension ratio and the engineering strain is defined by:

$$e = \frac{L_2 - L_1}{L_1} = \frac{L_2}{L_1} - \frac{L_1}{L_1} = \lambda - 1 \tag{3.20}$$

Therefore, if the extension ratio value is 1 then there is no engineering strain, and no deformation occurred in the environment.

Green Strain

Green strain is the most applicable strain tensors which is defined in Lagrangian coordinates. The following equation defines it:

$$\varepsilon_G = \frac{1}{2}\left(\frac{L_2^2 - L_1^2}{L_1^2}\right) = \frac{1}{2}\left(\lambda^2 - 1\right) \tag{3.21}$$

Almansi Strain

Another strain tensor which is very applicable is Eulerian strain tensor or Almansi strain tensor which is defined as below:

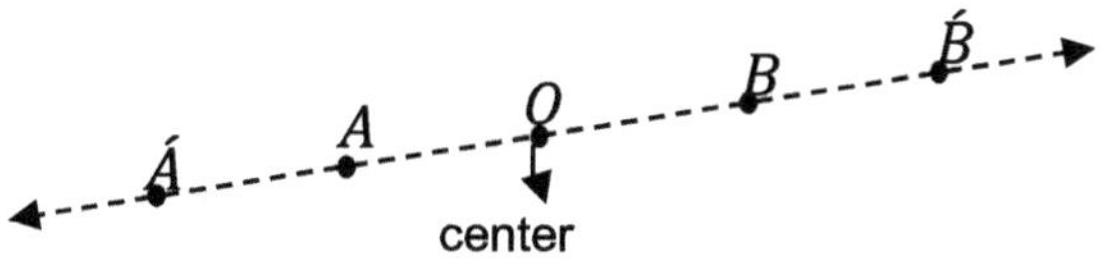

Fig. 3.2 The dilation of a line

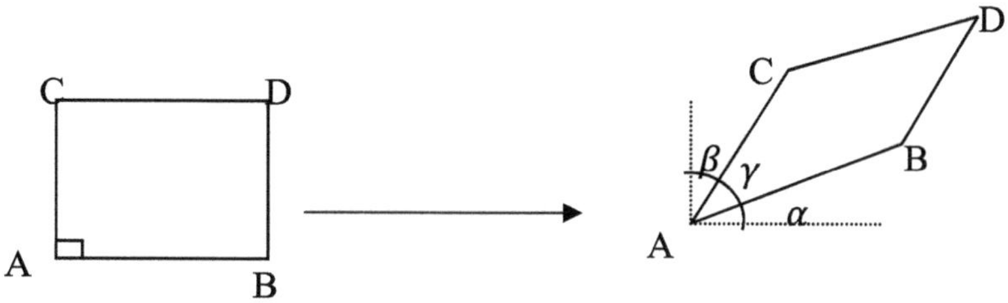

Fig. 3.3 The shear strain

$$\varepsilon_E = \frac{1}{2}\left(\frac{L_2^2 - L_1^2}{L_2^2}\right) = \frac{1}{2}\left(1 - \frac{1}{\lambda^2}\right) \tag{3.22}$$

The strains are either normal or shear strains. For example, when a continuum environment which obeys Hook law is under normal stress then normal strain has occurred which is referred to as dilation. The dilation is clear using Fig. 3.2. Note that dilation depends completely on the position or coordinates as well as the direction of the line in which the point is on the line.

In fact, the length of line AB is increased and dilated to line $A'B'$. It can be written $A'B' = \lambda$AB: where λ is the extension ratio.

We will discuss shear stress next, but shear stress is referred to the variation of the angle between AB and Ac lines as per below Fig. 3.3.

Now the total change in the initial angle, 90 degrees, is $\alpha + \beta$ for the square ABCD.

In this section, we obtain one of the most important tensors namely Green–Lagrange strain tensors, which has a wide application in theory deformation.

Figure 3.4 shows line AB in the environment R0 before deformation at $t = 0$ and its deformed configuration as $A'B'$ in environment R at $t = t0$.

We consider the coordinates of A, A', B, B' as below and assume these points are very close to each other:

$$A(X_1, X_2, X_3), \quad A'(x_1, x_2, x_3)$$

$$B(X_1 + dX_1, X_2 + dX_2, X_3 + dX_3), B'(x_1 + dx_1, x_2 + dx_2, x_3 + dx_3)$$

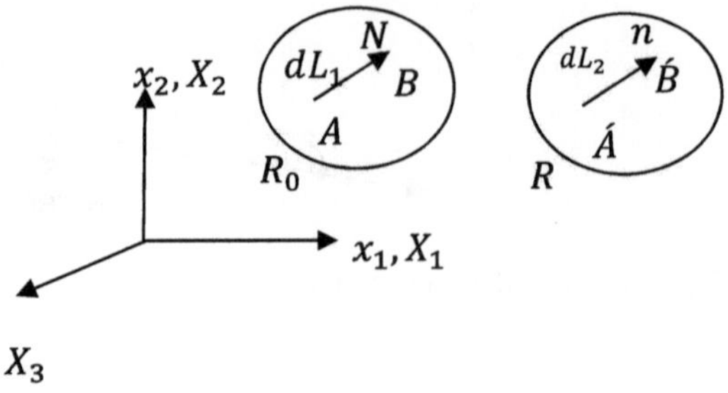

Fig. 3.4 The change in length of a linear element

$$AB\,length \rightarrow AB = \sqrt{dX_1^2 + dX_2^2 + dX_3^2} = dL_1$$
$$\acute{A}\acute{B}\,length \rightarrow \acute{A}\acute{B} = \sqrt{dx_1^2 + dx_2^2 + dx_3^2} = dL_2 \tag{3.23}$$

As we remember from mathematics, every vector that its magnitude is one called unit vector and any vector that is in the same direction as the unit vector is called the direction vector. For example, if n is the vector direction of a vector like $\vec{a}$:

$$\vec{n} = \frac{\vec{a}}{|a|} \tag{3.24}$$

Now we calculate the direction vectors (n, N) of $\overrightarrow{AB}$ and $\overrightarrow{\acute{A}\acute{B}}$ as below:

$$\vec{N} = \frac{dX_1 i_1 + dX_2 i_2 + dX_3 i_3}{\sqrt{dX_1^2 + dX_2^2 + dX_3^2}} \equiv \frac{dX_1 i_1 + dX_2 i_2 + dX_3 i_3}{dL_1}$$
$$\equiv \frac{dX_1}{dL_1} i_1 + \frac{dX_2}{dL_1} i_2 + \frac{dX_3}{dL_1} i_3$$

$$N_i = \frac{dX_i}{dL_1}\ n_i = \frac{dx_i}{dL_2} \tag{3.25}$$

Note to following mathematical operations as follows:

$$\begin{cases} dL_1^2 = dX_1^2 + dX_2^2 + dX_3^2 = dX_i.dX_i \\ dX_i = dX_1 i_1 + dX_2 i_2 + dX_3 i_3 \end{cases}$$

$$\begin{cases} dL_2^2 = dx_1^2 + dx_2^2 + dx_3^2 = dx_i.dx_i \\ dx_i = dx_1 i_1 + dx_2 i_2 + dx_3 i_3 \end{cases}$$

$$\begin{cases} dx_K = \frac{\partial x_k}{\partial X_j}.dX_j \\ x_i = u_i + X_i\left(\vec{x} = \vec{u} + \vec{X}\right) \end{cases} \rightarrow dx_k = \left(\delta_{kj} + \frac{\partial u_k}{\partial X_j}\right).dX_j \tag{3.26}$$

Relation (3.26) is the same as (3.14) because $F_{kj} = \frac{dx_k}{dX_j}$, Now we can write:

$$\begin{aligned} dx_k.dx_k &= \left[\delta K_i + \frac{\partial u_k}{\partial X_i}\right] dX_i.\left[\delta K_j + \frac{\partial u_k}{\partial X_j}\right] dX_j \\ &= \left[\delta K_i.\delta K_j + \delta K_i \frac{\partial u_k}{\partial X_j} + \delta K_j \frac{\partial u_k}{\partial X_i} + \frac{\partial u_k}{\partial X_i}.\frac{\partial u_k}{\partial X_j}\right] dX_i dX_j \\ &= \left[\delta_{ij} + \frac{\partial u_i}{\partial X_j} + \frac{\partial u_j}{\partial X_i} + \frac{\partial u_k}{\partial X_i}.\frac{\partial u_k}{\partial X_j}\right] dX_i dX_j \end{aligned}$$

$$d_{xk}.d_{xk} = \left[\delta_{ij} + u_{i,j} + u_{j,i} + u_{ki}.u_{k,j}\right] dX_i.dX_j \tag{3.27}$$

$$\lambda = \frac{dL_2}{\mathrm{dL}_1} \rightarrow \lambda^2 = \left(\frac{dL_2}{\mathrm{dL}_1}\right)^2$$

$$\begin{cases} dL_1^2 = dX_i.dX_i \\ dL_2^2 = dx_i.dx_i \end{cases} \rightarrow \lambda^2 = \left(\frac{dx_i.dx_i}{dX_i.dX_i}\right) \tag{3.28}$$

Noting relations (3.27) and (3.28), we find that λ^2 can be rewritten according to relation (3.27):

$$\begin{aligned} \lambda^2 &= \frac{dx_k.dx_k}{dX_i.dX_i} = \frac{dx_k.dx_k}{dL_1^2} \\ &= \frac{\left[\delta_{ij} + u_{i,j} + u_{j,i} + u_{ki}.u_{k,j}\right] dX_i.dX_j}{dL_1^2} \\ \lambda^2 &= \frac{\left[\delta_{ij} + u_{i,j} + u_{j,i} + u_{ki}.u_{k,j}\right] dX_i.dX_j}{\mathrm{dL}_1.\mathrm{dL}_1} \\ &\equiv \left[\delta_{ij} + u_{i,j} + u_{j,i} + u_{ki}.u_{k,j}\right].\frac{dX_i}{\mathrm{dL}_1}.\frac{dX_j}{\mathrm{dL}_1} \\ &= \left[\delta_{ij} + u_{i,j} + u_{j,i} + u_{ki}.u_{k,j}\right] N_i.N_j \end{aligned} \tag{3.29}$$

Relation (3.29) gives the exact magnitude of dilation for a element of length and expresses the ratio of the dilated length (dL_2) to the initial length (dL_1). Now, we express again the definition of Green–Lagrange strain:

$$\varepsilon_G = \frac{1}{2}(\lambda^2 - 1)$$

$$\lambda^2 = \left[\delta_{ij} + u_{i,j} + u_{j,i} + u_{ki} \cdot u_{k,j}\right] N_i \cdot N_j$$

$$\varepsilon_{\mathrm{G}} = \frac{1}{2}\left[\delta_{ij} \cdot N_i \cdot N_j + \left(u_{i,j} + u_{j,i} + u_{ki} \cdot u_{k,j}\right) \cdot N_i \cdot N_j - 1\right]$$

$$\delta_{ij} \cdot N_i \cdot N_j = \delta_{11} \cdot N_1 \cdot N_1 + \delta_{22} \cdot N_2 \cdot N_2 + \delta_{33} \cdot N_3 \cdot N_3 + \delta_{12} \cdot N_1 \cdot N_2 + \delta_{13} \cdot N_1 \cdot N_3 + \cdots$$

$$\begin{cases} N_1 \cdot N_1 + N_2 \cdot N_2 + N_3 \cdot N_3 = N_1^2 + N_2^2 + N_3^2 \equiv \frac{dX_1^2+dX_2^2+dX_3^2}{\left(\sqrt{dx_1^2+dx_2^2+dx_3^2}\right)^2} = 1 \\ \delta_{11} = \delta_{22} = \delta_{33} = 1 \end{cases}$$

$$\vec{N} = N_1 i_1 + N_2 i_2 + N_3 i_3 =$$

$$= \frac{dX_1 i_1}{\sqrt{dx_1^2 + dx_2^2 + dx_3^2}} + \frac{dX_2 i_2}{\sqrt{dx_1^2 + dx_2^2 + dx_3^2}} + \frac{dX_3 i_3}{\sqrt{dx_1^2 + dx_2^2 + dx_3^2}}$$

$$\delta_{12} = \delta_{13} = \ldots = 0 \rightarrow \delta_{12}N_1N_2 = \delta_{13}N_1N_3 = \ldots = 0$$

Finally:

$$\begin{aligned} \varepsilon_G &= \frac{1}{2}\left[1 + \left(u_{i,j} + u_{j,i} + u_{k,i}.u_{k,j}\right)N_i.N_j - 1\right] \\ &= \frac{1}{2}\left(u_{i,j} + u_{j,i} + u_{k,i}.u_{k,j}\right)N_i.N_j \end{aligned} \tag{3.30}$$

Relation (3.31) is called Green–Lagrange strain tensor and represents the components of strain for a continuum environment. The tensor is symmetrical and has nine components:

$$\begin{aligned} \varepsilon_{ij} &= \begin{bmatrix} \varepsilon_{11} & \varepsilon_{12} & \varepsilon_{13} \\ \varepsilon_{21} & \varepsilon_{22} & \varepsilon_{23} \\ \varepsilon_{31} & \varepsilon_{32} & \varepsilon_{33} \end{bmatrix} = \frac{1}{2}\left(u_{i,j} + u_{j,i} + u_{k,i}.u_{k,j}\right) \\ &= \frac{1}{2}\left(\frac{\partial u_i}{\partial X_j} + \frac{\partial u_j}{\partial X_i} + \frac{\partial u_k}{\partial X_i}.\frac{\partial u_k}{\partial X_j}\right) \end{aligned} \tag{3.31}$$

$\varepsilon_{11}, \varepsilon_{22}, \varepsilon_{33}$ are normal strains or axial strains which are in directions of Cartesian axes X_1, X_2, X_3 and $\varepsilon_{12} = \varepsilon_{21}$, $\varepsilon_{23} = \varepsilon_{32}$, $\varepsilon_{13} = \varepsilon_{31}$ are shear strains.

Now, we combine relations (3.30) and (3.31):

$$\varepsilon_G = \varepsilon_{ij}.N_i.N_j \tag{3.32}$$

In many engineering problems, we face small strains and deformations therefore, we can neglect the value of non-linear sentence in Green–Lagrange tensor $u_{k,i} \cdot u_{k,j}$ (it has very small value) and the engineering strain tensor is given by:

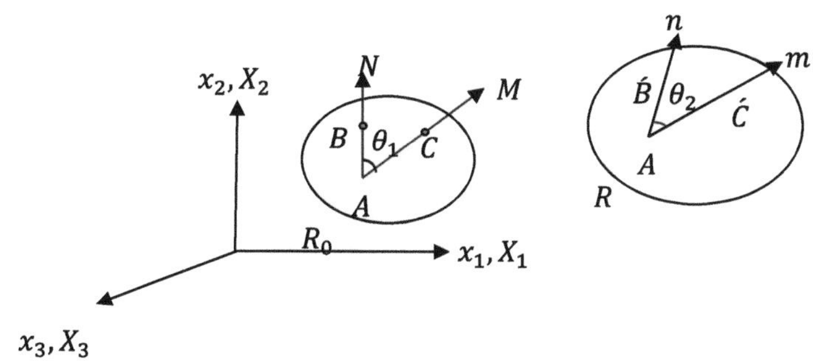

Fig. 3.5 The change of the angle between two lines due to deformation

$$\varepsilon_{ij} = \frac{1}{2}\left(u_{i,j} + u_{j,i}\right) = \frac{1}{2}\left(\frac{\partial u_i}{\partial X_j} + \frac{\partial u_j}{\partial X_i}\right) \tag{3.33}$$

Almansi strain tensor defines the strain in Eulerian coordinates as below.

$$E_{i,j} = \frac{1}{2}\left(\frac{\partial u_i}{\partial x_j} + \frac{\partial u_j}{\partial x_i} - \frac{\partial u_k}{\partial x_i}.\frac{\partial u_k}{\partial x_j}\right) \tag{3.34}$$

The procedure of obtaining Almansi tensor is the same as Green–Lagrange tensor exactly and can be done as practice by readers.

The Change of the Angle Between Two Lines After Deformation

We aim to study the change in value of an angle, between two lines, after deformation of the continuum environment. Figure 3.5 shows the lines AB and AC in R0 with the angle θ_1 and the lines are changed to $A\acute{B}, A\acute{C}$ after deformation in new environment R with the angle θ_2.

The direction vectors of the lines AB, AC, $\acute{B}$, $A\acute{C}$ are expressed the same as before. For example, if the length of AB is dL and length of $A\acute{B}$ as dl then:

$$N_i = \frac{dx_i}{dL}, n_i = \frac{dx_i}{l}$$

$$dx_i = \frac{\partial x_i}{\partial X_j}.dX_j \rightarrow \frac{dx_i}{dl} \times \frac{dl}{dL} = \frac{\partial x_i}{\partial X_j}.\frac{dX_j}{dL}$$

$$\rightarrow n_i.\frac{dl}{dl} = \frac{\partial x_i}{\partial X_j}.N_j \tag{3.35}$$

$$\varepsilon_G = \frac{1}{2}\left[\left(\frac{dl}{dL}\right)^2 - 1\right] \rightarrow \frac{dl}{dL} = \sqrt{1 + 2\varepsilon_G} \tag{3.36}$$

Combining relations (3.35) and (3.36):

$$n_i = \frac{1}{\sqrt{1+2\varepsilon_G(N)}}.\frac{\partial x_i}{\partial X_j}.N_j \tag{3.37}$$

Similarly for m and N directions:

$$m_i = \frac{1}{\sqrt{1+2\varepsilon_G(M)}}.\frac{\partial x_i}{\partial X_K}.M_K \tag{3.38}$$

The angles θ_1 and θ_2 can be defined as per dot product definition:

$$\begin{cases} N_j.M_k = |N_j|.|M_k|.Cos\theta_1 = Cos\theta_1 \\ n_i.m_i = |n_i|.|m_i|.Cos2 = Cos\theta_2 \end{cases}$$

We know:

$$\begin{aligned} |N_i| &= |M_k| \\ &= |n_i|.|m_i| = 1 \end{aligned}$$

Considering relations (3.37) and (3.38):

$$n_i.m_i = \left(\frac{1}{\sqrt{1+2\varepsilon_G(N)}}.\frac{\partial x_i}{\partial X_j}.N_j\right)\left(\frac{1}{\sqrt{1+2\varepsilon_G(M)}}.\frac{\partial x_i}{\partial X_K}.M_K\right) \tag{3.39}$$

$$\rightarrow (n_i.m_i).\sqrt{1+2\varepsilon_G(N)}\left(\sqrt{1+2\varepsilon_G(M)}\right)$$

$$\equiv \frac{\partial x_i}{\partial X_j}.\frac{\partial x_i}{\partial X_K}.N_j.M_k$$

$$\begin{cases} N_j.M_k = Cos\theta_1 \\ n_i.m_i = Cos\theta_2 \end{cases}$$

$$\rightarrow \sqrt{1+2\varepsilon_G(N)}\left(\sqrt{1+2\varepsilon_G(M)}\right).Cos\theta_2$$

$$= \frac{\partial x_i}{\partial X_j}.\frac{\partial x_i}{\partial X_K} \tag{3.40}$$

The right side of relation (3.40) is rewritten as:

$$\begin{cases} \partial x_i = \frac{dx_i}{dX_j}.\partial X_j \\ \partial x_i = \frac{dx_i}{dX_k}.\partial X_k \end{cases} \rightarrow \begin{cases} \frac{\partial x_i}{\partial X_j} = \frac{dx_i}{dX_j} \\ \frac{\partial x_i}{\partial X_k} = \frac{dx_i}{dX_k} \end{cases}$$

$$\frac{\partial x_i}{\partial X_j}.\frac{\partial x_i}{\partial X_k}(N_j)(M_k) = \frac{dx_i}{dX_j}.\frac{dx_i}{dX_k}.N_j.M_k$$

$$\begin{cases} N_j = \frac{dX_j}{dL} \\ M_k = \frac{dX_k}{dL} \\ dx_i.dx_i = (dl)^2 \end{cases} \rightarrow \frac{\partial x_i}{\partial X_j}.\frac{\partial x_i}{\partial X_k}.(N_j).(M_k) \equiv \frac{dx_i}{dX_j}.\frac{dx_i}{dX_k}.\frac{dX_j}{dL}.\frac{dX_k}{dL}$$

$$= \frac{dx_i.dx_i}{(dL)^2} \equiv \frac{(dl)^2}{(dL)^2} = \left(\frac{dl}{dL}\right)^2$$

If we note to relation (3.29):

$$\lambda^2 = \left(\frac{dl}{dL}\right)^2 = \left[\delta_{ij} + u_{i,j} + u_{j,i} + u_{ki}.u_{k,j}\right]N_i.N_j$$

We expand relation (3.50) noting to relation (3.29):

$$\begin{aligned} &\frac{\partial \mathrm{x_i}}{\partial \mathrm{X_j}}.\frac{\partial \mathrm{x_i}}{\partial \mathrm{X_k}}(N_j)(M_k) = \left(\frac{dl}{dL}\right)^2 \\ &= \left[\delta_{ij} + u_{i,j} + u_{j,i} + u_{ki}.u_{k,j}\right]N_i.M_j \\ &= \delta_{ij}.N_i.M_j + \left[u_{i,j} + u_{j,i} + u_{ki}.u_{k,j}\right]N_i.M_j \\ &= N_i.M_j + \left[u_{i,j} + u_{j,i} + u_{ki}.u_{k,j}\right]N_i.M_j \\ &\equiv Cos\theta_1 + 2\varepsilon_{ij}.N_i.M_j \end{aligned} \tag{3.41}$$

Note that based on Green–Lagrange strain tensor:

$$\varepsilon_{ij} = \frac{1}{2}\left[u_{i,j} + u_{j,i} + u_{k,i}.u_{k,j}\right]$$

Finally, we write the following equation based on relations (3.40) and (3.41):

$$\sqrt{1 + 2\varepsilon_G(N)}(\sqrt{1 + 2\varepsilon_G(M)}.Cos\theta_2 = Cos\theta_1 + 2\varepsilon_{ij}.N_i.M_j \tag{3.42}$$

Relation (3.42) shows the exact relation between the angle between two lines before (θ_1) and after deformation (θ_2).

Now, we intend to introduce the relation between engineering shear strain tensors, for very small strains, and components of Green–Lagrange strain tensors. According to Fig. 3.6, we assume Cartesian coordinates in two-dimensional plane $X_1 - \mathrm{X}_2$ with the center of origin O which the direction vectors are M and N:

The direction vectors are M (0,1,0) and N (1.0.0):

$$\begin{aligned} \theta_2 &= \frac{\pi}{2} - (\alpha + \beta) = \frac{\pi}{2} - \gamma \\ \rightarrow \cos\theta_2 &= \mathrm{Cos}\left(\frac{\pi}{2} - \gamma\right) = \sin\gamma \simeq \gamma \end{aligned}$$

On the other hand. $\alpha + \beta = \gamma$ has a very small value, therefore we can approximately write:

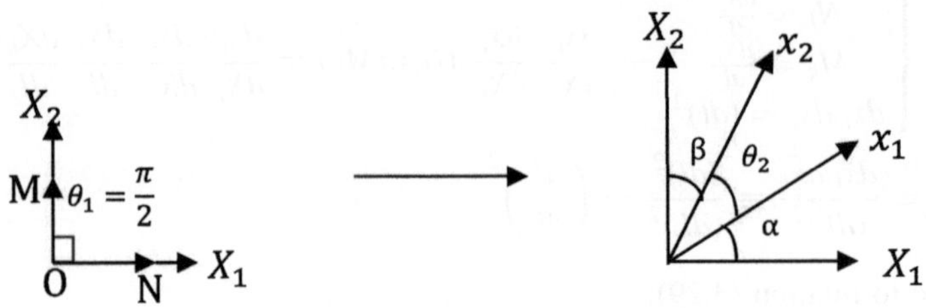

Fig. 3.6 The change of the angle between the axes of two-dimensional Cartesian coordinates for a small deformation

$$\begin{matrix} dx_1 \simeq dX_1 \\ dx_2 \simeq dX_2 \end{matrix} \rightarrow \frac{\mathrm{dl}}{\mathrm{dL}} = \sqrt{1 + 2\varepsilon_{\mathrm{G}}}$$

$$\rightarrow \frac{\mathrm{dl}}{\mathrm{dL}} \simeq 1 \rightarrow 1 = \sqrt{1 + 2\varepsilon_{\mathrm{G}}}$$

Noting the relation (3.42):

$$\sqrt{1 + 2\varepsilon_{\mathrm{G}}(\mathrm{N})}.\sqrt{1 + 2\varepsilon_{\mathrm{G}}(\mathrm{M})}.Cos\theta_2 = Cos\theta_1 + 2\varepsilon_{ij}.N_i.M_j$$

$$\begin{cases} \sqrt{1 + 2\varepsilon_{\mathrm{G}}} = 1 \rightarrow 1 \times 1 \times Cos\theta_2 = Cos\frac{\pi}{2} + 2\varepsilon_{ij}.N_i.M_j \\ \theta_1 = \frac{\pi}{2} \\ Cos\theta_2 = \gamma \rightarrow Cos\theta_2 = 2\varepsilon_{ij}.N_i.M_j \rightarrow \gamma = 2\varepsilon_{ij}N_iM_j \end{cases}$$

$$\gamma_{ij} = 2\varepsilon_{ij}.N_iM_j \tag{3.43}$$

Now, we calculate γ_{12} and γ_{13}:

$$\gamma_{12} = 2\varepsilon_{ij}N_1M_2$$

$N_1(1, 0, 0)$ The direction of axis X1:
The direction of axis X2: $M_2(0, 1, 0)$

$$\gamma_{12} = 2\begin{bmatrix} 1 & 0 & 0 \end{bmatrix} \times \begin{bmatrix} \varepsilon_{11} & \varepsilon_{12} & \varepsilon_{13} \\ \varepsilon_{21} & \varepsilon_{22} & \varepsilon_{23} \\ \varepsilon_{31} & \varepsilon_{32} & \varepsilon_{33} \end{bmatrix} \times \begin{bmatrix} 0 \\ 1 \\ 0 \end{bmatrix}$$

$$= 2\begin{bmatrix} \varepsilon_{11} & \varepsilon_{12} & \varepsilon_{13} \end{bmatrix}\begin{bmatrix} 0 \\ 1 \\ 0 \end{bmatrix} = 2\varepsilon_{12}$$

It means that the value of the shear engineering strain or (γ_{12}) is twice the corresponding component of Green–Lagrange strain tensor (ε_{12}).

Now, we choose X1and X3 axes with corresponding direction vectors N (1,0,0) and M (0,0,1), then we write for γ_{13}:

$$\gamma_{13} = 2\varepsilon_{ij}N_iM_j = 2\begin{bmatrix} 1 & 0 & 0 \end{bmatrix}\begin{bmatrix} \varepsilon_{11} & \varepsilon_{12} & \varepsilon_{13} \\ \varepsilon_{21} & \varepsilon_{22} & \varepsilon_{23} \\ \varepsilon_{31} & \varepsilon_{32} & \varepsilon_{33} \end{bmatrix}\begin{bmatrix} 0 \\ 0 \\ 1 \end{bmatrix}$$

$$\equiv 2\begin{bmatrix} \varepsilon_{11} & \varepsilon_{12} & \varepsilon_{13} \end{bmatrix}\begin{bmatrix} 0 \\ 0 \\ 1 \end{bmatrix} = 2\varepsilon_{13}$$

In fact, we can write:

$$\gamma_{ij} = 2\varepsilon_{ij}(i \neq j) \tag{3.44}$$

Relation (3.44) represents the relationship between the engineering shear strain and the components of Green–Lagrange strain tensor.

The Rotation Tensor

The rotation tensor is defined based on the gradient of displacement vector or $\frac{\partial u_i}{\partial X_j}$:

$$\frac{\partial u_i}{\partial X_j} = \begin{bmatrix} \frac{\partial u_1}{\partial X_1} & \frac{\partial u_1}{\partial X_2} & \frac{\partial u_1}{\partial X_3} \\ \frac{\partial u_2}{\partial X_1} & \frac{\partial u_2}{\partial X_2} & \frac{\partial u_2}{\partial X_3} \\ \frac{\partial u_3}{\partial X_1} & \frac{\partial u_3}{\partial X_2} & \frac{\partial u_3}{\partial X_3} \end{bmatrix} \tag{3.45}$$

This tensor can be decomposed into two other tensors as:

$$\frac{\partial u_i}{\partial X_j} = \frac{1}{2}\left(\frac{\partial u_i}{\partial X_j} + \frac{\partial u_j}{\partial X_i}\right) + \frac{1}{2}\left(\frac{\partial u_i}{\partial X_j} - \frac{\partial u_j}{\partial X_i}\right)$$

Or:

$$e_{ij} = \frac{1}{2}\left(\frac{\partial u_i}{\partial X_j} + \frac{\partial u_j}{\partial X_i}\right)$$

$$w_{ij} = \frac{1}{2}\left(\frac{\partial u_i}{\partial X_j} - \frac{\partial u_j}{\partial X_i}\right) \tag{3.46}$$

e_{ij} is the Green–Lagrange strain tensor for small strains or engineering strain tensor. w_{ij} is called the spin tensor.

Principal Strains

The same as the procedure which was used to obtain the principal stress and related principal directions, we once again use it for calculating the principal strain and the directions. If the principal strains are defined as ε_p then they are calculated by multiplying normal vector n_j and strain tensor ε_{ij}:

$$\varepsilon_p = \varepsilon_{ij}.n_j \tag{3.47}$$

If λ exists as a scalar that we can define $\lambda.n_i = \varepsilon_p$ then:

$$\varepsilon_p = \lambda n_i = \lambda\delta_{ij}n_j = \varepsilon_{ij}.n_j \tag{3.48}$$

$$\rightarrow (\varepsilon_{ij} - \lambda\delta_{ij})n_j = 0 \tag{3.49}$$

The Eq. (3.49) has nonzero answers if the following determinant is zero:

$$\left|\varepsilon_{ij} - \lambda\delta_{ij}\right| = 0 \rightarrow$$

$$\begin{cases} (\varepsilon_{11} - \lambda)n_1 + \varepsilon_{12}n_2 + \varepsilon_{13}n_3 = 0 \\ \varepsilon_{21}n_1 + (\varepsilon_{22} - \lambda)n_2 + \varepsilon_{23}n_3 = 0 \\ \varepsilon_{31}n_3 + \varepsilon_{32}n_2 + (\varepsilon_{33} - \lambda)n_3 = 0 \end{cases}$$

$$\rightarrow \lambda^3 - \lambda^2 I_1 + \lambda I_2 - I_3 = 0$$

$$I_1 = \varepsilon_{ii} = \varepsilon_{11} + \varepsilon_{22} + \varepsilon_{33}$$

$$I_2 = \frac{1}{2}\left[\varepsilon_{ii}.\varepsilon_{jj} - \varepsilon_{ij}.\varepsilon_{ji}\right] = \varepsilon_{11}.\varepsilon_{22} + \varepsilon_{22}\varepsilon_{33} + \varepsilon_{33}\varepsilon_{11} - \varepsilon_{12}^2 - \varepsilon_{23}^2 - \varepsilon_{31}^2$$

$$I_3 = \left|\varepsilon_{ij}\right|$$

The equation has three real answers indicating principal strains and we can use relation (3.48) to obtain the principal directions in terms of $\lambda_1, \lambda_2, \lambda_3$.

The deviator strain tensor and spherical strain tensor

Like the stress tensor, each strain tensor can be decomposed into a deviator $\overline{\varepsilon_{ij}}$ and spherical strain tensor ε_m:

Average strain $\varepsilon_m = \frac{\varepsilon_{11}+\varepsilon_{22}+\varepsilon_{33}}{3} \rightarrow \varepsilon_{ij} = \varepsilon_m.\delta_{ij} + \overline{\varepsilon_{ij}}$

$$\overline{\varepsilon_{ij}} = \varepsilon_{ij} - \varepsilon_m.\delta_{ij} = \begin{bmatrix} \varepsilon_{11} & \varepsilon_{12} & \varepsilon_{13} \\ \varepsilon_{21} & \varepsilon_{22} & \varepsilon_{23} \\ \varepsilon_{31} & \varepsilon_{32} & \varepsilon_{33} \end{bmatrix} - \begin{bmatrix} \varepsilon_m & 0 & 0 \\ 0 & \varepsilon_m & 0 \\ 0 & 0 & \varepsilon_m \end{bmatrix} \tag{3.50}$$

The Octahedral Strain

Assume the principal strain tensor are defined as below tensor along with the principal directions:

$$\varepsilon_p = \begin{bmatrix} \varepsilon_{11} & 0 & 0 \\ 0 & \varepsilon_{22} & 0 \\ 0 & 0 & \varepsilon_{33} \end{bmatrix}$$

Similar to octahedral planes in stress, there are planes that make equal angles with respect to principal strain directions, and we call them octahedral planes where the related strain vectors are named octahedral strain vectors. The number of the planes is 8 and their normal vectors are:

$$\frac{1}{\sqrt{3}}(-1,-1,-1),\ \frac{1}{\sqrt{3}}(\pm 1,1,-1),\ \frac{1}{\sqrt{3}}(\pm 1,-1,1)$$
$$\frac{1}{\sqrt{3}}(-1,1,1),\ \frac{1}{\sqrt{3}}(\pm 1,1,1)$$

The normal component of octahedral strain and tangential (shear) component of octahedral strain can be calculated the same as stress part:

$$\varepsilon_N \text{oct} = \frac{\varepsilon_{11}+\varepsilon_{22}+\varepsilon_{33}}{3}$$

$$\varepsilon_S \text{oct} = \frac{2}{3}\sqrt{(\varepsilon_{11}-\varepsilon_{22})^2+(\varepsilon_{22}-\varepsilon_{33})^2+(\varepsilon_{33}-\varepsilon_{11})^2} \tag{3.51}$$

The Volume Strain

There are two methods for calculating the volume strain of a continuum environment:

1. Using Strain Definition Concepts:

Based on the Lagrangian strain definition for unit of volume:

$$\varepsilon_v = \frac{1}{2}\left[\left(\frac{dv_2}{dv_1}\right)^2 - 1\right]$$

dv_2 is the volume of deformed element and dv_1 is the initial volume of the element, therefore as per relation (3.36):

$$\varepsilon_G = \frac{1}{2}\left[\left(\frac{dl}{dL}\right)^2 - 1\right] \rightarrow \frac{dl}{dL} = \sqrt{1+2\varepsilon_G}$$

For three i_1, i_2, i_3 directions, we can write the following based on relation (3.36):

$$\frac{dx_1}{dX_1} = \sqrt{1+2\varepsilon_{G_1}} = \sqrt{1+2\varepsilon_{11}}$$

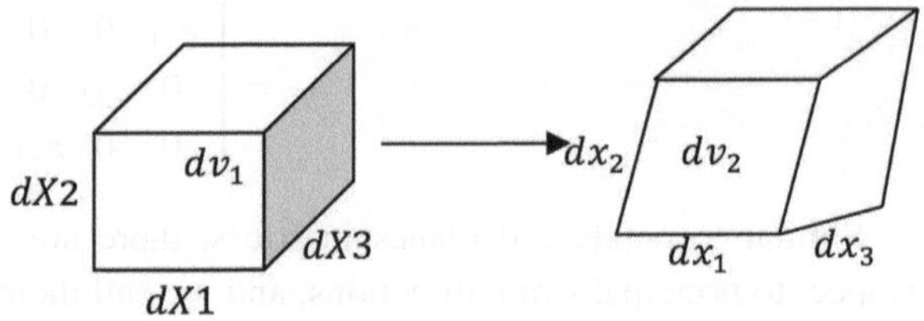

Fig. 3.7 The deformation of a cubic element

$$\frac{dx_2}{dX_2} = \sqrt{1 + 2\varepsilon_{G_2}} = \sqrt{1 + 2\varepsilon_{22}}$$
$$\frac{dx_3}{dX_3} = \sqrt{1 + 2\varepsilon_{G_3}} = \sqrt{1 + 2\varepsilon_{33}}$$

$$\frac{dx_i}{dX_i} = \sqrt{1 + 2\varepsilon_{ii}} \tag{3.52}$$

Now, the ratio of the deformed volume to initial is rewritten as:

$$\frac{dv_2}{dv_1} = \left(\frac{dx_1}{dX_1}\right)\left(\frac{dx_2}{dX_2}\right)\left(\frac{dx_3}{dX_3}\right)$$

where dx_i and dX_i are the dimensions of a rectangular cube before and after deformation. Therefore, the volume of the cube is calculated by:

$$\left(\frac{dv_2}{dv_1}\right)^2 = \left(\frac{dx_1}{dX_1}\right)^2\left(\frac{dx_2}{dX_2}\right)^2\left(\frac{dx_3}{dX_3}\right)^2$$
$$= \left(\sqrt{1 + 2\varepsilon_{11}}\right)^2\left(\sqrt{1 + 2\varepsilon_{11}}\right)^2\left(\sqrt{1 + 2\varepsilon_{11}}\right)^2 \tag{3.53}$$

We substitute relation (3.53) in the volume strain relation:

$$\varepsilon_{\text{v}} = \frac{1}{2}[(1 + 2\varepsilon_{11})(1 + 2\varepsilon_{22})(1 + 2\varepsilon_{33}) - 1] \tag{3.54}$$

When the occurred deformations are very small, then relation (3.54) is expressed below by deleting the non-linear parts:

$$\varepsilon_{\text{v}} = \varepsilon_{11} + \varepsilon_{22} + \varepsilon_{33} \tag{3.55}$$

2. Using Concept of Deformation of a Continuum Environment:

Figure 3.7 shows a cubic element with dX_1, dX_2, dX_3 dimensions before deformation and the deformed shape is shown which is a parallelepiped with dx_1, dx_2, dx_3 dimensions.

The volume change of the initial cube and deformed parallelepiped is calculated using relation (3.37):

$$
\begin{aligned}
dv_1 &= dX_1.dX_2.dX_3 \\
dv_2 &= (dx_1 \times dx_2).dx_3
\end{aligned}
\tag{3.56}
$$

We remember the following relation also:

$$dx_k = \frac{\partial x_k}{\partial X_i} dX_i$$

Therefore:

$$dv_2 = (dx_1 \times dx_2).dx_3 = \left(\frac{\partial x_m}{\partial X_1}.i_m \times \frac{\partial x_n}{\partial X_2} i_n.\frac{\partial x_t}{\partial X_2} i_t\right).dX_1.dX_2.dX_3 \tag{3.57}$$

The displacement gradient tensor F was also known as:

$$dx_k = \frac{\partial x_k}{\partial X_i} dX_i = F.dX_i$$

Thus, each of deformed volume vectors which are the sides of the parallelepiped can be defined by:

$$
\begin{aligned}
d\vec{x}_1 &= F.d\vec{X}_1 = F.dX_1 \cdot i_1 \\
d\vec{x}_2 &= F.d\vec{X}_2 = F.dX_2 \cdot i_2 \\
d\vec{x}_3 &= F.d\vec{X}_3 = F.dX_3 \cdot i_3
\end{aligned}
$$

Note to the following operation and $\begin{bmatrix} 1 \\ 0 \\ 0 \end{bmatrix}$ is the equivalent matrix of unit vector i_1:

$$
\begin{aligned}
d\vec{x}_1 = F.dX_1.i_1 &= \begin{bmatrix} \frac{\partial x_1}{\partial X_1} & \frac{\partial x_1}{\partial X_2} & \frac{\partial x_1}{\partial X_3} \\ \frac{\partial x_2}{\partial X_1} & \frac{\partial x_2}{\partial X_2} & \frac{\partial x_2}{\partial X_3} \\ \frac{\partial x_3}{\partial X_1} & \frac{\partial x_3}{\partial X_2} & \frac{\partial x_3}{\partial X_3} \end{bmatrix} .dX_1. \begin{bmatrix} 1 \\ 0 \\ 0 \end{bmatrix} \\
&= \left(\frac{\partial x_1}{\partial X_1} + \frac{\partial x_2}{\partial X_1} + \frac{\partial x_3}{\partial X_1}\right) dX_1 = \left(\frac{\partial x_m}{\partial X_1}.i_m\right) dX_1
\end{aligned}
$$

The values of can be obtained as practice. Finally, the deformed volume is calculated by:

$$dv_2 = (d\vec{x}_1 \times d\vec{x}_2) \cdot d\vec{x}_3 = \left(\left(\frac{\partial x_m}{\partial X_1} \cdot i_m\right.\right.$$

$$\times \frac{\partial x_n}{\partial X_2} i_n\Big) \cdot \frac{\partial x_t}{\partial X_3} i_t\Big) \cdot dX_1 \cdot dX_2 \cdot dX_3$$

$$= \begin{vmatrix} \frac{\partial x_1}{\partial X_1} & \frac{\partial x_1}{\partial X_2} & \frac{\partial x_1}{\partial X_3} \\ \frac{\partial x_2}{\partial X_1} & \frac{\partial x_2}{\partial X_2} & \frac{\partial x_2}{\partial X_3} \\ \frac{\partial x_3}{\partial X_1} & \frac{\partial x_3}{\partial X_2} & \frac{\partial x_3}{\partial X_3} \end{vmatrix} \cdot dX_1 \cdot dX_2 \cdot dX_3 \tag{3.58}$$

It should be noted that dX_1, dX_2, dX_3 are the vectors in direction of Cartesian coordinates which give the volume of the cubic element according to existing formula for volume of a cube, but the deformed element is a parallelepiped with new vectors dx_1, dx_2, dx_3 which are deformed vectors.

$$dv_2 = \begin{vmatrix} \frac{\partial x_1}{\partial X_1} & \frac{\partial x_1}{\partial X_2} & \frac{\partial x_1}{\partial X_3} \\ \frac{\partial x_2}{\partial X_1} & \frac{\partial x_2}{\partial X_2} & \frac{\partial x_2}{\partial X_3} \\ \frac{\partial x_3}{\partial X_1} & \frac{\partial x_3}{\partial X_2} & \frac{\partial x_3}{\partial X_3} \end{vmatrix} .dX_1.dX_2.dX_3 \tag{3.59}$$

Relation (3.58) can be rewritten based on relation (3.17) and definition of Jacobien J or F_{ij}:

$$dV_2 = |\mathrm{J}|.dV_1 = |F_{\mathrm{ij}}|dV_1 \tag{3.60}$$

Relation (3.59) gives the relationship between a deformed volume and an initial one in a continuum environment. If J = 1, then there is no deformation or change of volume. In general, the main condition for deformation of a continuum environment is that J must be greater than zero.

The Change of Surface in Continuum Environment

Assume that continuum environment R0 is deformed and changed to new R. Now, we intend to calculate the change of surface during this motion. Therefore, we have a known surface element ds_1 in R0 and try to obtain unknown surface element ds_2 in R. Figure 3.8 shows the surface elements ds_1, ds_2 with initial dimensions of dL_1, dL_2 and dl_1, dl_2. We assume that normal vector of ds_1 is N and n for ds_2. We recall relation (3.35) for calculation of parallelepiped, furthermore, we also assume the vector N is changed to M after deformation. Now, a length vector dL is chosen in the direction of N, and another length vector dl is chosen in the direction of M. Using relation (3.35), we calculate the volume of the parallelepiped objects which are produced by the vectors (dl_1, dl_2, dl), (dL_1, dL_2, dL)

$$dV_1 = (dL_1 \times dL_2).dL = N.ds_1.dL$$
$$dV_1 = (dl_1 \times dl_2).dl = n.ds_2.dl$$

On the other hand, we know from relation (3.58): $dv_2 = |\mathrm{J}|.dv_1 dv_2 = |\mathrm{J}|.dv_1$

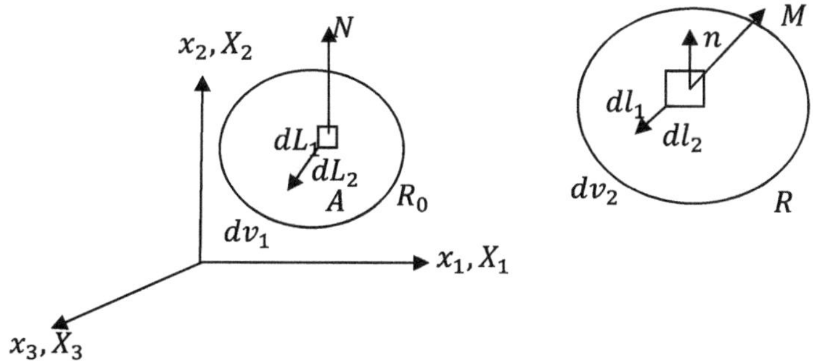

Fig. 3.8 The change of surface in a continuum environment

$$dv_2 = J.dv_1 \rightarrow nds_2dl = Nds_1.dL.|\mathrm{J}|$$

We mention again that all these relations are based on vectors, and we write them according to index notation:

$$n_\mathrm{m}.\mathrm{ds}_2.\mathrm{Mm.dl} = |J|.N_n.ds_1.N_n.dL$$

Also, we know that $N_\mathrm{n}.\mathrm{N_n} = 1$ based on unit vector dot product:

$$\rightarrow n_\mathrm{m}.\mathrm{ds}_2.\mathrm{Mm.dl} = |\mathrm{J}|.\mathrm{ds}_1.\mathrm{dL}$$

Using relation (3.35):

$$n_\mathrm{i}.\frac{\mathrm{dl}}{\mathrm{dL}} = \frac{\partial \mathrm{x_i}}{\partial \mathrm{X_j}}.\mathrm{N_j}$$

Again, $N_\mathrm{I}.\mathrm{N_I} = 1$:

$$\mathrm{n_m}.\mathrm{ds}_2.\frac{\partial \mathrm{x_r}}{\partial \mathrm{X_I}}.\mathrm{N_I}.\mathrm{dL} = |\mathrm{J}|.\mathrm{ds}_1.\mathrm{dL}$$

We multiple above relation by $\left[\frac{\partial x_r}{\partial X_I}\right]^{-1}.N_I$:

$$n_r.ds_2 = |J|.\left[\frac{\partial x_r}{\partial X_I}\right]^{-1}.N_I.ds_1 \tag{3.61}$$

Equation (3.60) expresses the relation between the surface changes of an element before and after deformation. There is a very important note that $\left[\frac{\partial \mathrm{x_r}}{\partial \mathrm{X_I}}\right]^{-1}$ is the inverse of displacement gradient tensor F and should be considered in transpose form in Eq. (3.60).

Rate of Deformation Tensor and Vorticity/spin Tensor

The velocity gradient is a tensor in Eulerian coordinates which is defined as:

$$\frac{\partial V_i}{\partial x_j} = \begin{bmatrix} \frac{\partial V_1}{\partial x_1} & \frac{\partial V_1}{\partial x_2} & \frac{\partial V_1}{\partial x_3} \\ \frac{\partial V_2}{\partial x_1} & \frac{\partial V_2}{\partial x_2} & \frac{\partial V_2}{\partial x_3} \\ \frac{\partial V_3}{\partial x_1} & \frac{\partial V_3}{\partial x_2} & \frac{\partial V_3}{\partial x_3} \end{bmatrix} \tag{3.62}$$

This tensor can be rewritten as:

$$\frac{\partial V_i}{\partial x_j} = \frac{1}{2}\left(\frac{\partial V_i}{\partial x_j} + \frac{\partial V_j}{\partial x_i}\right) + \frac{1}{2}\left(\frac{\partial V_i}{\partial x_j} - \frac{\partial V_j}{\partial x_i}\right)$$

where

$$\begin{aligned} D_{ij} &= \frac{1}{2}\left(\frac{\partial V_i}{\partial x_j} + \frac{\partial V_j}{\partial x_i}\right) \\ W_{ij} &= \frac{1}{2}\left(\frac{\partial V_i}{\partial x_j} - \frac{\partial V_j}{\partial x_i}\right) \end{aligned} \tag{3.63}$$

D_{ij} is the rate of deformation tensor and W_{ij} is vorticity or spin tensor. Note that vorticity tensor is not the same as rotation tensor that was defined in Lagrangian coordinates. Vorticity tensor expresses the rate of displacement in Eulerian coordinates.

Problems

1. For the given motion field in Cartesian coordinates, find:
   A. The deformation gradient tensor.
   B. Green–Lagrange Tensor.
   C. The ratio of deformed volume to initial one.
   D. The deformed surface vector if the initial surface vector before deformation is i_2

$$\begin{cases} x_1 = X_1 \\ x_2 = 3X_3 \\ x_3 = -2X_2 \end{cases}$$

$$\begin{cases} x_1 = X_1 \\ x_2 = 3X_3 \\ x_3 = -2X_2 \end{cases} \rightarrow^{x_i = u_i + X_i} \begin{cases} u_1 = 0 \\ u_2 = 3X_3 - X_2 \\ u_3 = -2X_2 - X_3 \end{cases}$$

For example:

$$\begin{cases} x_2 = 3X_3 \\ x_3 = u_2 + x_2 \rightarrow u_2 = x_2 - X_2 \end{cases}$$

$$\rightarrow u_2 = 3X_3 - X_2$$

A. $F = \left[\frac{\partial x_i}{\partial X_j}\right] = \begin{bmatrix} \frac{\partial x_1}{\partial X_1} & \frac{\partial x_1}{\partial X_2} & \frac{\partial x_1}{\partial X_3} \\ \frac{\partial x_2}{\partial X_1} & \frac{\partial x_2}{\partial X_2} & \frac{\partial x_2}{\partial X_3} \\ \frac{\partial x_3}{\partial X_1} & \frac{\partial x_3}{\partial X_2} & \frac{\partial x_3}{\partial X_3} \end{bmatrix} = \begin{bmatrix} 1 & 0 & 0 \\ 0 & 0 & 3 \\ 0 & -2 & 0 \end{bmatrix}$

For example:

$$\begin{cases} \frac{\partial x_1}{\partial X_1} = \frac{\partial (X_1)}{\partial X_1} = 1 \\ x_1 = X_1 \end{cases}$$

B.
$$\varepsilon_{ij} = \frac{1}{2}\left(u_{i,j} + u_{j,i} + u_{k,i}.u_{k,j}\right)$$
$$\varepsilon_{11} = \frac{1}{2}\left(\frac{\partial u_1}{\partial X_1} + \frac{\partial u_1}{\partial X_1} + \frac{\partial u_1}{\partial X_1}.\frac{\partial u_1}{\partial X_1} + \frac{\partial u_2}{\partial X_1}.\frac{\partial u_2}{\partial X_1} + \frac{\partial u_3}{\partial X_1}.\frac{\partial u_{13}}{\partial X_1}\right) = 0$$

For example:

$$\begin{cases} \frac{\partial u_1}{\partial X_1} = \frac{\partial (0)}{\partial X_1} = 0 \\ \frac{\partial u_2}{\partial X_1} = \frac{\partial (3X_3 - X_2)}{\partial X_1} = 0 \end{cases}$$

$$\varepsilon_{12} = \frac{1}{2}\left(\frac{\partial u_1}{\partial X_2} + \frac{\partial u_2}{\partial X_1} + \frac{\partial u_1}{\partial X_1}.\frac{\partial u_1}{\partial X_2} + \frac{\partial u_2}{\partial X_1}.\frac{\partial u_2}{\partial X_2} + \frac{\partial u_3}{\partial X_1}.\frac{\partial u_3}{\partial X_2}\right) = 0$$

$$\varepsilon_{13} = \frac{1}{2}\left(\frac{\partial u_1}{\partial X_3} + \frac{\partial u_3}{\partial X_1} + \frac{\partial u_1}{\partial X_1}.\frac{\partial u_1}{\partial X_3} + \frac{\partial u_2}{\partial X_1}.\frac{\partial u_2}{\partial X_3} + \frac{\partial u_3}{\partial X_1}.\frac{\partial u_3}{\partial X_3}\right) = 0$$

$$\varepsilon_{22} = \frac{1}{2}\left(\frac{\partial u_2}{\partial X_2} + \frac{\partial u_2}{\partial X_2} + \frac{\partial u_1}{\partial X_2}.\frac{\partial u_1}{\partial X_2} + \frac{\partial u_2}{\partial X_2}.\frac{\partial u_2}{\partial X_2} + \frac{\partial u_3}{\partial X_2}.\frac{\partial u_{13}}{\partial X_2}\right) =$$

$$\frac{1}{2}(-1 - 1 + 0 + (-1)(-1) + (-2)(- - 2) = \frac{3}{2}$$

$$\varepsilon_{23} = \frac{1}{2}\left(\frac{\partial u_2}{\partial X_3} + \frac{\partial u_3}{\partial X_2} + \frac{\partial u_1}{\partial X_2}.\frac{\partial u_1}{\partial X_3} + \frac{\partial u_2}{\partial X_2}.\frac{\partial u_2}{\partial X_3} + \frac{\partial u_3}{\partial X_2}.\frac{\partial u_{13}}{\partial X_3}\right) =$$

$$\frac{1}{2}(3 - 2 + 0 + (-1)(3) + (-2)(-1) = 0$$

$$\varepsilon_{33} = \frac{1}{2}\left(\frac{\partial u_3}{\partial X_3} + \frac{\partial u_3}{\partial X_2} + \frac{\partial u_1}{\partial X_3}.\frac{\partial u_1}{\partial X_3} + \frac{\partial u_2}{\partial X_3}.\frac{\partial u_2}{\partial X_3} + \frac{\partial u_3}{\partial X_3}.\frac{\partial u_{13}}{\partial X_3}\right) =$$

$$\frac{1}{2}(-1-1+(3)(3)+(-1)(-1)+(-1)(-1)=4$$

Green–Lagrange tensor:

$$\rightarrow \varepsilon_{\mathrm{ij}} = \begin{bmatrix} 0 & 0 & 0 \\ 0 & \frac{3}{2} & 0 \\ 0 & 0 & 4 \end{bmatrix}$$

C. $dV_2 = |J|.dV_1$

$$\rightarrow \frac{dV_2}{dV_1} = |J| = |F_{ij}| = \left|\left[\frac{\partial x_i}{\partial X_j}\right]\right| = \begin{vmatrix} 1 & 0 & 0 \\ 0 & 0 & 3 \\ 0 & -2 & 0 \end{vmatrix} = 6$$

Note that F_{ij} has been put from A.

D. $n_r.ds_2 = |J|.\left[\frac{\partial x_r}{\partial X_j}\right]^{-1}.N_1.ds_1$

$$\begin{cases} N_I = i_2 \rightarrow N_I = \begin{bmatrix} 0 \\ 1 \\ 0 \end{bmatrix} \\ ds_1 = 1 \\ |J| = 6 \end{cases}$$

$$\left[\frac{\partial x_r}{\partial X_j}\right] = \begin{bmatrix} 1 & 0 & 0 \\ 0 & 0 & 3 \\ 0 & -2 & 0 \end{bmatrix} \rightarrow \left[\frac{\partial x_r}{\partial X_j}\right]^{-1} = \begin{bmatrix} 1 & 0 & 0 \\ 0 & 0 & -\frac{1}{2} \\ 0 & \frac{1}{3} & 0 \end{bmatrix}$$

Now we put the transpose form of the above tensor in surface relation:

$$\rightarrow n_r.ds_2 = 6 \times \begin{bmatrix} 1 & 0 & 0 \\ 0 & 0 & \frac{1}{3} \\ 0 & -\frac{1}{2} & 0 \end{bmatrix}_{3\times 3} \times \begin{bmatrix} 0 \\ 1 \\ 0 \end{bmatrix}_{3\times 1} = 6\begin{bmatrix} 0 \\ 0 \\ -\frac{1}{2} \end{bmatrix} = -3i_3$$

2. The deformation of a continuum body is given as below:

$$\begin{cases} x_1 = KX_2 + X_1 \\ x_2 = X_2 \\ x_3 = X_3 \end{cases}$$

A. Find the extension of the element along the i_1 direction.
B. Find the extension of the element along the $i_1 + i_2$ direction.
C. Find the angle between directions i_1, i_2 after deformation.

A. First we calculate u_i according to $x_i = u_i + X_i$:

$$\begin{cases} x_1 = u_1 + X_1 \\ x_2 = u_2 + X_2 \\ x_3 = u_3 + X_3 \end{cases} \rightarrow \begin{cases} x_1 = X_1 + KX_2 = u_1 + X_1 \\ \rightarrow u_1 = KX_2 \\ u_2 = u_3 = 0 \end{cases}$$

$$\varepsilon_g = \varepsilon_{ij}.N_i.N_j$$

$$\begin{aligned} &\rightarrow \vec{N} = i_1 \rightarrow N_1 = 1,\ N_2 = N_3 = 0 \\ &= N_1 i_1 + N_2 i_2 + N_3 i_3 \end{aligned}$$

$$\varepsilon_g^{(i_1)} = \varepsilon_{11}.N_1.N_1 \rightarrow \varepsilon_{g\,(i_1)} = \varepsilon_{11} \times 1 \times 1 = \varepsilon_{11}$$

$$\varepsilon_{11} = \frac{1}{2}\left(\frac{\partial u_1}{\partial X_1} + \frac{\partial u_1}{\partial X_1} + \frac{\partial u_1}{\partial X_1}.\frac{\partial u_1}{\partial X_1} + \frac{\partial u_2}{\partial X_1}.\frac{\partial u_2}{\partial X_1} + \frac{\partial u_3}{\partial X_1}.\frac{\partial u_3}{\partial X_1}\right) = 0$$

$$\rightarrow \varepsilon_g(i_1) = 0,\ 2\varepsilon_g(i_1) = \lambda^2 - 1 = \varepsilon_{11} = 0$$

$$\rightarrow \lambda^2 - 1 = 0 \rightarrow \lambda = 1$$

B. $\lambda^2 = \left(\frac{dL_2}{dL_1}\right)^2 = \left(\delta_{ij} + u_{i,j} + u_{j,i} + u_{k,i}.u_{k,j}\right)N_i.N_j$

$$i_1 + i_2 \rightarrow \vec{N} = \frac{i_1 + i_2}{\sqrt{1+1}} = \frac{1}{\sqrt{2}}i_1 + \frac{1}{\sqrt{2}}i_2 = N_1 i_1 + N_2 i_2$$

$$\rightarrow \begin{cases} N_1 = N_2 = \frac{1}{\sqrt{2}} \\ N_3 = 0 \end{cases}$$

$$\begin{aligned} \lambda^2 &= \left(\delta_{11} + u_{1,1} + u_{1,1} + u_{k,1}.u_{k,1}\right)N_1.N_1 \\ &+ \left(\delta_{12} + u_{1,2} + u_{2,1} + u_{k,1}.u_{k,2}\right)N_1.N_2 \end{aligned}$$

$$+\left(\delta_{21}+u_{2,1}+u_{1,2}+u_{k,2}.u_{k,1}\right)N_2.N_1$$
$$+\left(\delta_{22}+u_{2,2}+u_{2,2}+u_{k,2}.u_{k,2}\right)N_2.N_2$$

Note that $u_{1,3}$, $u_{2,3}$ are set zero because $N_3=0$.

$$\rightarrow \lambda^2=(1+0+0+0)\left(\frac{1}{\sqrt{2}}\right)\left(\frac{1}{\sqrt{2}}\right)+(0+k+0)\frac{1}{\sqrt{2}}\times\frac{1}{\sqrt{2}}$$
$$+(0+k)\left(\frac{1}{\sqrt{2}}\times\frac{1}{\sqrt{2}}\right)+\left(1+0+k^2\right)\frac{1}{\sqrt{2}}.\frac{1}{\sqrt{2}}$$

$$\rightarrow \lambda^2=\frac{1}{2}+\frac{k}{2}+\frac{k}{2}+\frac{1}{2}+\frac{k^2}{2}=\frac{k^2}{2}+k+1=\frac{k^2+2k+2}{2}$$

$$\rightarrow \lambda_{i_1+i_2}=\sqrt{\frac{k^2+2k+2}{2}}$$

C. See the schematic of the change in angle between i_1 and i_2

$$\sqrt{1+2\varepsilon_g(i_1)}.\sqrt{1+2\varepsilon_g(i_2)}.\cos\varphi=\frac{\pi}{2}+2\varepsilon_{ij}.N_i.M_j$$

$$N_1 \rightarrow' i_1 \rightarrow \vec{N_1}=i_1=N_1 i_1$$

$$N_2 \rightarrow i_2 \rightarrow \vec{N_2}=i_2=\vec{M}=M_2 i_2$$

$$\varepsilon_g(i_1)=\varepsilon_{11}.N_1.N_1=0$$

$$\varepsilon_g(i_2)=\varepsilon_{22}.M_2.M_2=\frac{k^2}{2}\times 1\times 1=\frac{k^2}{2}$$

Note that other components including N_2 N_3, M_1, M_3 are zero both in $\varepsilon_g(i_1)$ and $\varepsilon_g(i_2)$. Thus. Only $\varepsilon_{11}.N_1.N_1$ and $\varepsilon_{22}.M_2.M_2$ are not zero.

$$\varepsilon_{22}=\frac{1}{2}\left(\frac{\partial u_2}{\partial X_2}+\frac{\partial u_2}{\partial X_2}+\frac{\partial u_1}{\partial X_2}.\frac{\partial u_1}{\partial X_2}+\frac{\partial u_2}{\partial X_2}.\frac{\partial u_2}{\partial X_2}+\frac{\partial u_3}{\partial X_2}.\frac{\partial u_{13}}{\partial X_3}\right)$$

$$= \frac{1}{2}(0 + 0 + \mathrm{K.K} + 0 + 0) = \frac{\mathrm{K}^2}{2}$$

$$\rightarrow \sqrt{1 + 2(0)}.\sqrt{1 + 2\left(\frac{K^2}{2}\right)}.Cos\varphi = 0 + 2\varepsilon_{12}N_1M_2$$

Note when we expand $2\varepsilon_{ij}.N_i.M_j$ then all sentences except are zero because only $\mathrm{M}_2 = \mathrm{N}_1 = 1$ and $\mathrm{N}_2 = \mathrm{M}_1 = \mathrm{N}_3 = \mathrm{M}_3 = 0$. By $\varepsilon_{12} = \frac{\mathrm{k}}{2}$ we have:

$$\sqrt{1 + 0}.\sqrt{1 + \mathrm{k}^2}.\mathrm{Cos}\varphi = 2\left(\frac{\mathrm{k}}{2}\right)(1)(1)$$

$$\rightarrow Cos\varphi = \frac{K}{\sqrt{1 + k^2}}$$

and in the final 3.The position of particle is initially at (X_1, X_2, X_3) position given below:

$$\begin{cases} x_1 = X_1 + (X_1 + X_2)t \\ x_2 = X_2 + (X_1 + X_2)t \\ x_3 = X_3 \end{cases}$$

A. Find the velocity of the particle at t = 2 s that was initially in the position (1,0,0)
B. Find the velocity of the particle that is in the position (1,1,0) at t = 2 s.

$$(X_1, X_2, X_3) \rightarrow (1, 1, 0)$$

$$x_1 = \mathrm{X}_1 + (\mathrm{X}_1 + \mathrm{X}_2)\mathrm{t} \rightarrow \overset{\circ}{\mathrm{x}}_1 = \mathrm{V}_1 = (\mathrm{X}_1 + \mathrm{X}_2) = 1 + 1 = 2\mathrm{m/s}$$

A. $$x_2 = \mathrm{X}_2 + (\mathrm{X}_1 + \mathrm{X}_2)\mathrm{t} \rightarrow \overset{\circ}{\mathrm{x}}_2 = \mathrm{V}_2 = (\mathrm{X}_1 + \mathrm{X}_2) = 1 + 1 = 2\mathrm{m/s}$$

$$x_3 = \mathrm{X}_3 \rightarrow \overset{\circ}{\mathrm{x}}_3 = 0$$

$$\rightarrow V = 2i_1 + 2i_2 m/s$$

$$\begin{cases} x_1 = X_1 + (X_1 + X_2)t \\ x_2 = X_2 + (X_1 + X_2)t \rightarrow \\ x_3 = X_3 \end{cases}$$

$$\begin{cases} X_1 = \frac{x_1+(1+t)-x_2t}{1+2t} \\ X_2 = \frac{-x_1t+x_2(1+t)}{1+2t} \\ X_3 = x_3 \end{cases}$$

B. $$\begin{cases} V_1 = X_1 + X_2 = \frac{x_1+(1+t)-x_2t}{1+2t} + \frac{-x_1t+x_2(1+t)}{1+2t} \\ V_2 = X_1 + X_2 = \frac{x_1+(1+t)-x_2t}{1+2t} + \frac{-x_1t+x_2(1+t)}{1+2t} \\ V_{3=0} \end{cases}$$

$$\begin{cases} V_1 = V_2 = \frac{x_1+x_2}{1+2t} = \frac{1+1}{1+2(2)} = \frac{2}{5} \\ V_3 = 0 \\ (1,1,0) \rightarrow (x_1, x_2, 0) \end{cases}$$

$t = 2^s$

$$\rightarrow \vec{V} = \frac{2}{5}\mathrm{i}_1 + \frac{2}{5}\mathrm{i}_2$$

4. A square is deformed by the displacement function given below. First find the components of the displacement functions (u_1, u_2, u_3) and then Green–Lagrange strain tensor.

$$u_i = C_1 + C_2X_1 + C_3X_2 + C_4X_1X_2$$

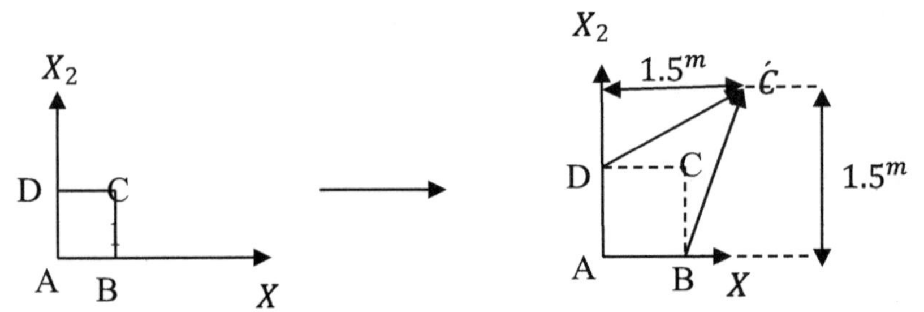

As it can be seen, only corner C has been deformed and other corners were unchanged. Therefore, we consider the initial coordinates and deformed one for each corner and calculate displacement functions u_1, u_2, u_3

$$A(0,0)\textit{before deformation} \rightarrow X_1 = X_2 = 0$$
$$A(0,0)\text{after deformation} \rightarrow x_1 = x_2 = 0$$

$$\begin{cases} u_1 = C_1 + C_2X_1 + C_3X_2 + C_4X_1X_2 \\ u_2 = C_1 + C_2X_1 + C_3X_2 + C_4X_1X_2 \end{cases}$$

$$X_1 = \mathrm{X}_2 = 0 \rightarrow \begin{cases} u_1 = C_1 \\ u_2 = C_2 \end{cases}$$

But we know:

$$\begin{cases} x_1 = u_1 + X_1 \\ x_2 = u_2 + X_2 \end{cases} \rightarrow \begin{matrix} 0 = u_1 + 0 \\ 0 = u_2 + 0 \end{matrix} \rightarrow u_1 = u_2 = 0$$

$C_1 = \mathrm{C}_2 = 0$ Therefore:

$$B(0,0)\text{before deformation} \rightarrow (X_1, X_2) = (1, 0)$$
$$B(0,0)\text{after deformation} \rightarrow (x_1, x_2) = (1, 0)$$

$$\begin{cases} x_1 = u_1 + X_1 \\ x_2 = u_2 + X_2 \end{cases} \rightarrow \begin{matrix} 1 = u_1 + 1 \\ 0 = u_2 + 0 \end{matrix} \rightarrow u_1 = 0, u_2 = 0$$

$$\begin{cases} u_1 = C_1 + C_2X_1 + C_3X_2 + C_4X_1X_2 \rightarrow u_1 = 0 = C_2 \\ u_2 = C_1 + C_2X_1 + C_3X_2 + C_4X_1X_2 \rightarrow u_2 = 0 = C_2 \end{cases} \rightarrow \mathrm{C}_2 = 0$$

$$\begin{cases} X_1 = x_1 = 1 \\ X_2 = x_2 = 0 \end{cases}$$

$$\begin{cases} u_1 = 0 \\ u_2 = 0 \end{cases}$$

$$D(0, 1)\text{before deformation} \rightarrow (X_1, X_2) = (0, 1)$$

$$D(0, 1)\text{after deformation} \rightarrow (x_1, x_2) = (0, 1)$$

$$\rightarrow \begin{cases} x_1 = u_1 + X_1 \\ x_2 = u_2 + X_2 \end{cases} \rightarrow \begin{matrix} 0 = u_1 + 0 \\ 0 = u_2 + 0 \end{matrix} \rightarrow u_1 = u_2 = 0$$

$$\begin{cases} u_1 = C_3X_2 + C_4X_1X_2 \rightarrow u_1 = 0 = C_3 \\ u_2 = C_3X_2 + C_4X_1X_2 \rightarrow u_2 = 0 = C_3 \rightarrow C_3 = 0 \\ X_1 = x_1 = 1 \\ X_2 = x_2 = 0 \\ u_1 = u_2 = 0 \\ X_1 = 0, x_1 = 0 \\ X_1 = 1, x_1 = 1 \end{cases}$$

$$C(1,1)\text{before deformation} \rightarrow (X_1, X_2) = (1,1)$$
$$\acute{C}(1.5,1.5)\text{after deformation} \rightarrow (x_1, x_2) = (1.5,1.5)$$

$$\rightarrow \begin{cases} x_1 = u_1 + X_1 \\ x_2 = u_2 + X_2 \end{cases} \rightarrow \begin{matrix} 1.5 = u_1 + 1 \\ 1.5 = u_2 + 1 \end{matrix} \rightarrow u_1 = u_2 = 0.5$$

$$\rightarrow \begin{cases} u_1 = C_4 X_1 X_2 \\ u_2 = C_4 X_1 X_2 \end{cases} \rightarrow C_4 = 0.5$$

$$\begin{cases} u_1 = u_2 = 0.5 \\ X_1 = X_2 = 1 \\ x_1 = x_2 = 1.5 \end{cases}$$

Therefore, the displacement functions are:

$$\begin{cases} u_1 = \frac{1}{2} X_1 X_2 \\ u_2 = \frac{1}{2} X_1 X_2 \end{cases}$$

$$\begin{cases} u_1 = u_2 = \frac{1}{2} X_1 X_2 \\ \varepsilon_{ij} = \frac{1}{2}\left(u_{i,j} + u_{j,i} + u_{k,i}.u_{k,j}\right) \end{cases}$$

$$\varepsilon_{11} = \frac{1}{2}\left(\frac{\partial u_1}{\partial X_1} + \frac{\partial u_1}{\partial X_1} + \frac{\partial u_1}{\partial X_1}.\frac{\partial u_1}{\partial X_1} + \frac{\partial u_2}{\partial X_1}.\frac{\partial u_2}{\partial X_1} + \frac{\partial u_3}{\partial X_1}.\frac{\partial u_3}{\partial X_1}\right)$$
$$= \frac{1}{2}\left(\frac{1}{2}X_2 + \frac{1}{2}X_2 + \frac{1}{2}X_2.\frac{1}{2}X_2 + \frac{1}{2}X_2.\frac{1}{2}X_2 + 0\right) = \frac{1}{2}X_2 + \frac{1}{4}X_2^2$$

$$\varepsilon_{22} = \frac{1}{2}X_1 + \frac{1}{4}X_1^2$$

$$\varepsilon_{12} = \frac{1}{4}(X_1 + X_2) + \frac{1}{4}(X_1X_2 + X_1X_2) \equiv \frac{1}{4}(X_1 + X_2) + \frac{X_1X_2}{2}$$

$$\varepsilon_{13} = \varepsilon_{23} = \varepsilon_{33} = 0$$

$$\rightarrow \varepsilon_{ij} = \begin{bmatrix} \frac{X_2}{2} + \frac{X_2^2}{4} & \frac{1}{4}(X_1 + X_2) + \frac{X_1X_2}{2} & 0 \\ \frac{1}{4}(X_1 + X_2) + \frac{X_1X_2}{2} & \frac{X_1}{2} + \frac{X_1^2}{4} & 0 \\ 0 & 0 & 0 \end{bmatrix}$$

5. The displacement field is given for a cube with side of unit as below. Find the Increase of AB after deformation.

$$\begin{cases} u_1 = 2X_1 \\ u_2 = u_3 = 0 \end{cases}$$

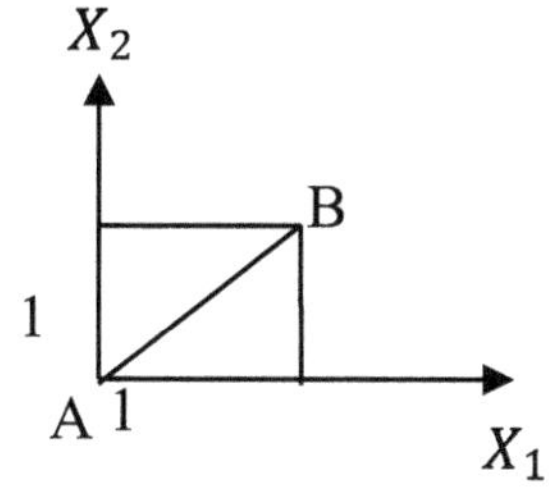

Ab is considered as a vector $\overrightarrow{AB}$ and the direction vector N is written:

$$\begin{matrix} A(0,0) \\ B(1,1) \end{matrix} \rightarrow \overrightarrow{AB}(1,1) \rightarrow \vec{N}_{AB} = \frac{1}{\sqrt{1+1}} i_1 + \frac{1}{\sqrt{1+1}} i_2 = \frac{i_1}{\sqrt{2}} + \frac{i_2}{\sqrt{2}}$$

First, the extension ration (λ) is calculated and then the strain magnitude e is calculated using $e = \lambda - 1e = \lambda - 1$

$$\lambda^2 = \left(\delta_{ij} + \frac{\partial u_i}{\partial X_j} + \frac{\partial u_j}{\partial X_i} + \frac{\partial u_k}{\partial X_i} \cdot \frac{\partial u_k}{\partial X_j}\right) N_i . N_j$$

$$\begin{aligned} N_3 = 0 \rightarrow \lambda^2 = & \left(\delta_{11} + u_{1,1} + u_{1,1} + u_{k,1}.u_{k,1}\right) \\ & + \left(\delta_{12} + u_{1,2} + u_{2,1} + u_{k,1}.u_{k,2}\right) N_1.N_2 \\ & + \left(\delta_{12} + u_{1,2} + u_{2,1} + u_{k,1}.u_{k,2}\right) N_1.N_2 \\ & + \left(\delta_{21} + u_{2,1} + u_{1,2} + u_{k,2}.u_{k,1}\right) N_2.N_1 \\ & + \left(\delta_{22} + u_{1,2} + u_{2,1} + u_{k,1}.u_{k,2}\right) N_1.N_2 \end{aligned}$$

$$u_{k,1}.u_{k,1} = \frac{\partial u_1}{\partial X_1} \cdot \frac{\partial u_1}{\partial X_1} + \frac{\partial u_2}{\partial X_1} \cdot \frac{\partial u_2}{\partial X_1} + \frac{\partial u_3}{\partial X_1} \cdot \frac{\partial u_3}{\partial X_1} = 4 + 0 + 0 = 4$$

$$u_{k,2}.u_{k,1} = 0,\ u_{k,2}.u_{k,2} = 0, \quad \delta_{11} = \delta_{22} = 1$$

$$\begin{aligned} \rightarrow \lambda^2 = & (1 + 2 + 2 + 4)\left(\frac{1}{\sqrt{2}}\right)\left(\frac{1}{\sqrt{2}}\right) + (1 + 0)\left(\frac{1}{\sqrt{2}}\right)\left(\frac{1}{\sqrt{2}}\right) \\ & + 2(0)\left(\frac{1}{\sqrt{2}}\right)\left(\frac{1}{\sqrt{2}}\right) = 5 \end{aligned}$$

Note that only $u_{1,1} = \frac{\partial u_1}{\partial X_1} = 2$ and other sentences like.
$u_{1,2} = \frac{\partial u_1}{\partial X_2}$ and $u_{2,2} = \frac{\partial u_2}{\partial X_2}$ are zero.

$$e = \lambda - 1 = \sqrt{5} - 1 \rightarrow \frac{L_2}{L_1} = e = \sqrt{5} - 1$$

The initial length of Ab is $L_1 = \sqrt{1+1}$ and its increased length is $L_2 = \sqrt{2}(\sqrt{5} - 1)^m L_2 = \sqrt{2}(\sqrt{5} - 1)^m$

6. The velocity field is defined below in Eulerian coordinate and time dependent. Find the rate of deformation tensor (D_{ij}) and vorticity tensor (w_{ij}) for the point (1, −1, 0) at t = 1 s

$$\begin{cases} V_1 = 2tx_1 sinx_3 \\ V_2 = 2tx_2 Cosx_3 \\ V_3 = 0 \end{cases}$$

$$\begin{cases} t = 1^s \\ (x_1, x_2, x_3) = (1, -1, 0) \end{cases}$$

$$D_{ij} = \frac{1}{2}\left(\frac{\partial V_i}{\partial x_j} + \frac{\partial V_j}{\partial x_i}\right)$$

$$\rightarrow \begin{cases} D_{11} = \frac{\partial V_1}{\partial x_1} = 2t \sin x_3 \\ D_{12} = \frac{1}{2}\left(\frac{\partial V_1}{\partial x_2} + \frac{\partial V_2}{\partial x_1}\right) = \frac{1}{2}(0+0) = 0 \\ D_{13} = \frac{1}{2}\left(\frac{\partial V_1}{\partial x_3} + \frac{\partial V_3}{\partial x_1}\right) = \frac{1}{2}(2tx_1 Cosx_3) = tx_1 Cosx_3 = 1 \times 1 \times Cos0 = 1 \end{cases}$$

$$D_{21} = \frac{1}{2}\left(\frac{\partial V_2}{\partial x_1} + \frac{\partial V_1}{\partial x_2}\right) = 0$$

$$D_{22} = \frac{1}{2}\left(\frac{\partial V_2}{\partial x_2} + \frac{\partial V_2}{\partial x_2}\right) = 2tcosx_3 = 2$$

$$D_{23} = \frac{1}{2}\left(\frac{\partial V_2}{\partial x_3} + \frac{\partial V_3}{\partial x_2}\right) = 0$$

$$D_{31} = \frac{1}{2}\left(\frac{\partial V_3}{\partial x_1} + \frac{\partial V_1}{\partial x_3}\right) = tx_1 cosx_3 = 1$$

$$D_{32} = \frac{1}{2}\left(\frac{\partial V_3}{\partial x_2} + \frac{\partial V_2}{\partial x_3}\right) = 0$$

$$D_{33} = \frac{1}{2}\left(\frac{\partial V_3}{\partial x_3} + \frac{\partial V_3}{\partial x_3}\right) = 0$$

$$\rightarrow D_{ij} = \begin{bmatrix} 0\,0\,1 \\ 1\,2\,0 \\ 1\,0\,0 \end{bmatrix}$$

$$w_{ij} = \frac{1}{2}\left(\frac{\partial V_i}{\partial x_j} - \frac{\partial V_j}{\partial x_i}\right)$$

$$w_{11} = \frac{1}{2}\left(\frac{\partial V_1}{\partial x_1} - \frac{\partial V_1}{\partial x_1}\right) = 0$$
$$w_{22} = \frac{1}{2}\left(\frac{\partial V_2}{\partial x_2} - \frac{\partial V_2}{\partial x_2}\right) = 0$$
$$w_{33} = \frac{1}{2}\left(\frac{\partial V_3}{\partial x_3} - \frac{\partial V_3}{\partial x_3}\right) = 0$$
$$w_{12} = \frac{1}{2}\left(\frac{\partial V_1}{\partial x_1} - \frac{\partial V_2}{\partial x_1}\right) = \frac{1}{2}(0 - 0) = 0$$
$$w_{13} = \frac{1}{2}\left(\frac{\partial V_1}{\partial x_3} - \frac{\partial V_3}{\partial x_1}\right) = \frac{1}{2}(2tx_1cosx_3) = tx_1cosx_3 = 1 \times 1cos0 = 1$$
$$w_{31} = \frac{1}{2}\left(\frac{\partial V_3}{\partial x_1} - \frac{\partial V_1}{\partial x_3}\right) = \frac{1}{2}(-2tx_1cosx_3) = -1$$
$$w_{23} = w_{32} = 0 \rightarrow w_{ij} = \begin{bmatrix} 0 & 0 & 1 \\ 0 & 0 & 0 \\ -1 & 0 & 0 \end{bmatrix}$$

7. For the displacement field given below, find the ration of deformed volume to initial volume.

$$\begin{cases} u_1 = \frac{1}{4}(X_3 - X_2) \\ u_2 = \frac{1}{4}(X_1 - X_3) \\ u_1 = \frac{1}{4}(X_2 - X_1) \end{cases}$$

$$F_{ij} = J = \begin{bmatrix} 1 + \frac{\partial u_1}{\partial x_1} & \frac{\partial u_1}{\partial x_2} & \frac{\partial u_1}{\partial x_3} \\ \frac{\partial u_2}{\partial x_1} & 1 + \frac{\partial u_2}{\partial x_2} & \frac{\partial u_2}{\partial x_3} \\ \frac{\partial u_3}{\partial x_1} & \frac{\partial u_3}{\partial x_2} & 1 + \frac{\partial u_3}{\partial x_3} \end{bmatrix} = \begin{bmatrix} 1 & \frac{1}{4} & \frac{-1}{4} \\ \frac{1}{4} & 1 & -\frac{1}{4} \\ -\frac{1}{4} & \frac{1}{4} & 1 \end{bmatrix}$$

$$\rightarrow [J] = \begin{bmatrix} 1 & \frac{1}{4} & -\frac{1}{4} \\ \frac{1}{4} & 1 & -\frac{1}{4} \\ -\frac{1}{4} & \frac{1}{4} & 1 \end{bmatrix} = \frac{57}{64}$$

$$\frac{V_2}{V_1} = |J| = \frac{57}{64}$$

8. For below given motion field in Lagrangian coordinates, find Almansi Strain tensor when the strains are very small.

$$\begin{cases} x_1 = X_1 + X_2 \\ x_2 = X_2 + X_3 \\ x_3 = X_3 + X_1 \end{cases}$$

First we obtain X_1, X_2, and X_3 in terms of x_1, x_2, and x_3.

$$\begin{cases} X_1 = \frac{x_1 - x_2 + x_3}{2} \\ X_2 = \frac{x_1 - x_3 + x_2}{2} \\ X_3 = \frac{-x_1 + x_2 + x_3}{2} \end{cases}$$

Now the displacement vector components are calculated:

$$\begin{matrix} x_1 = X_1 + u_1 \rightarrow \\ x_1 = X_2 + u_2 \rightarrow \\ x_1 = X_3 + u_3 \rightarrow \end{matrix} \begin{cases} u_1 = x_1 - X_1 = (X_1 + X_2) - X_1 = X_2 \\ u_2 = x_2 - X_2 = (X_2 + X_2) - X_2 = X_3 \\ u_3 = x_3 - X_3 = (X_1 + X_2) - X_3 = X_1 \end{cases}$$

$$\begin{cases} u_1 = X_2 = \frac{x_1 - x_3 + x_2}{2} \\ u_2 = X_3 = \frac{-x_1 + x_2 + x_3}{2} \\ u_3 = X_1 = \frac{x_1 - x_2 + x_3}{2} \end{cases}$$

For small strains, the Almansi strain tensor is:

$$E_{ij} = \frac{1}{2}\left(\frac{\partial u_i}{\partial x_j} + \frac{\partial u_j}{\partial x_i}\right)$$

$$E_{11} = \frac{1}{2}\left(\frac{\partial u_1}{\partial x_1} + \frac{\partial u_1}{\partial x_1}\right) = \frac{\partial u_1}{\partial x_1} = \frac{1}{2}$$

$$E_{12} = \frac{1}{2}\left(\frac{\partial u_1}{\partial x_2} + \frac{\partial u_2}{\partial x_1}\right) = \frac{1}{2}\left(\frac{1}{2} - \frac{1}{2}\right) = 0$$

$$E_{13} = \frac{1}{2}\left(\frac{\partial u_1}{\partial x_3} + \frac{\partial u_3}{\partial x_1}\right) = \frac{1}{2}\left(\frac{-1}{2} + \frac{1}{2}\right) = 0$$

$$E_{21} = E_{12} = 0, E_{31} = E_{13} = 0$$

$$E_{22} = \frac{1}{2}\left(\frac{\partial u_2}{\partial x_2} + \frac{\partial u_2}{\partial x_2}\right) = \frac{\partial u_2}{\partial x_2} = \frac{1}{2}$$

$$E_{23} = E_{32} = \frac{1}{2}\left(\frac{\partial u_2}{\partial x_3} + \frac{\partial u_3}{\partial x_2}\right) = \frac{1}{2}\left(\frac{1}{2} + \frac{-1}{2}\right) = 0$$

$$E_{33} = \frac{1}{2}\left(\frac{\partial u_3}{\partial x_3} + \frac{\partial u_3}{\partial x_3}\right) = \frac{\partial u_3}{\partial x_3} = \frac{1}{2}$$

$$\rightarrow E_{ij} = \begin{bmatrix} \frac{1}{2} & 0 & 0 \\ 0 & \frac{1}{2} & 0 \\ 0 & 0 & \frac{1}{2} \end{bmatrix}$$

9. Find the Green–Lagrange strain tensor and spin tensor and then obtain the magnitude of vorticity vector at point (0, 2, −1)

$$\begin{cases} u_1 = (X_1 - X_3)^2 \\ u_2 = (X_2 + X_3)^2 \\ u_1 = -X_1 X_2 \end{cases}$$

$$\frac{\partial \mathrm{u_i}}{\partial \mathrm{Xj}} = \begin{bmatrix} \frac{\partial u_1}{\partial X_1} & \frac{\partial u_1}{\partial X_2} & \frac{\partial u_1}{\partial X_3} \\ \frac{\partial u_2}{\partial X_1} & \frac{\partial u_2}{\partial X_2} & \frac{\partial u_2}{\partial X_3} \\ \frac{\partial u_3}{\partial X_1} & \frac{\partial u_3}{\partial X_2} & \frac{\partial u_3}{\partial X_3} \end{bmatrix} = \begin{bmatrix} 2(X_1 - X_3) & 0 & -2(X_1 - X_3) \\ 0 & 2(X_1 + X_3) & 2(X_2 + X_3) \\ -X_2 & -X_1 & 0 \end{bmatrix}$$

$$\varepsilon_{\mathrm{ij}} = \frac{1}{2}\left(\mathrm{u_{i,j}} + \mathrm{u_{j,i}}\right) = \begin{bmatrix} 2(X_1 - X_3) & 0 & -(X_1 - X_3) - \frac{X_2}{2} \\ 0 & 2(X_2 + X_3) & (X_2 + X_3) - \frac{X_1}{2} \\ -(X_1 - X_3) - \frac{X_2}{2} & (X_2 + X_3) - \frac{X_1}{2} & 0 \end{bmatrix}$$

$$\rightarrow \varepsilon_{\mathrm{ij}} = \begin{bmatrix} 2 & 0 & -2 \\ 0 & 2 & 1 \\ -2 & 1 & 0 \end{bmatrix}$$

$$w_{ij} = \frac{1}{2}(u_{i,j} + u_{j,i}) \text{ or } w_{ij} = \frac{\partial u_i}{\partial Xj} - \varepsilon_{ij}$$

$$\rightarrow w_{ij} = \begin{bmatrix} 0 & 0 & -(X_1 - X_3) + \frac{X_2}{2} \\ 0 & 0 & X_2 + X_3 + \frac{X_1}{2} \\ (X_1 - X_3) - \frac{X_2}{2} & -(X_2 + X_3) - \frac{X_1}{2} & 0 \end{bmatrix}$$

$$\xrightarrow{(0,\,2,\,-1)} w_{ij} = \begin{bmatrix} 0 & 0 & 0 \\ 0 & 0 & 1 \\ 0 & -1 & 0 \end{bmatrix}$$

The Constitutive Laws in Continuum Mechanics

4

Most of the classical physics laws are defined based on the conservation of some quantitative physical properties. These laws include mass, electrical charge, energy, and are not limited to momentum. The laws are based on general concepts and are not limited to specific materials, which are called universal physical laws. They have been expressed by mathematical formulation, which shall be satisfied in any physical phenomenon. It should be noted that the second law of thermodynamics is different from other conservation laws and is defined by inequality.

Mass Conservation Law

We first define the mass conservation law in Lagrangian coordinates. An initial frame reference R0 with volume V_0 and mass m is considered. According to the mass conservation law, the material inside the continuum environment R_0 remains constant after deformation. Therefore, if the initial density of the frame reference is ϱ_0 and the density after deformation is ϱ then we have:

$$\varrho_0 = \lim_{\delta V_0 \to 0} \frac{\delta m}{\delta V_0} \quad \varrho = \lim_{\delta V \to 0} \frac{\delta m}{\delta V} \tag{4.1}$$

V is the volume of the deformed environment R. We remember the deformation gradient relation once again:

$$F_{ij} = \frac{\partial x_i}{\partial X_j}, \ |F_{ij}| = \left| \frac{\partial x_i}{\partial X_j} \right| = |J|$$

Noting to relation (4.1), we see:

K. Hassani and F. Azimifar, *Introduction to Continuum Mechanics*, Synthesis Lectures on Mechanical Engineering, https://doi.org/10.1007/978-3-032-05076-2_4

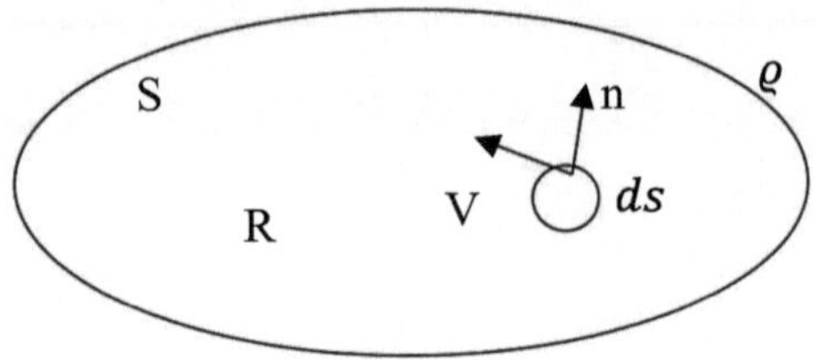

Fig. 4.1 The continuum environment R

$$\delta m(constant\ always) \to \varrho_0.\delta V_0 = \varrho.\delta V \to \frac{\varrho_0}{\varrho} = \frac{dV}{dV_0} \tag{4.2}$$

In the last chapter, we showed that the deformed volume of the continuum environment and its initial volume are related by the Jacobian or $|J|$

$$dV = |J|dV_0 \to \frac{dV}{dV_0} = |J|$$

We compare the above relation to relation (4.2) and the result:

$$\frac{\varrho_0}{\varrho} = |J| \tag{4.3}$$

Relation (4.3) expresses the mass conservation law in Lagrangian coordinates. We note that when $J = 0$, then the density of the frame reference is zero, or the deformed density is infinite. Therefore, J is always assumed $J \neq 0$.

Now, we try to define the mass conservation law in Eulerian coordinates because the law should be applied in many other applications based on the velocity components. Figure 4.1 presents an environment with surface S. The Mass conservation law indicates that the increase of mass inside R is equal to the rate of input mass to R through S borders. The rate of mass flow per area element d_s is calculated by multiplying the density by d_s and the normal velocity vector component. According to the definition of the mass conservation law:

The increase of mass inside = Rate of input mass to R

$$\to \iiint_V \frac{\partial \varrho}{\partial t} dV = -\iint_s \varrho.\vec{V}.n.ds \tag{4.4}$$

Note that the normal component of the velocity vector is determined by the dot product of the vector with the normal vector of the surface S.

Moreover, $\frac{\partial \varrho}{\partial t}$ is the rate of density increase for a certain point inside R. The minus sign in relation (4.4) is because $\varrho.V.n.ds$ is negative, as the direction of the normal vector n is toward the outside of the surface S. Thus, to show that the whole mass input rate to R is positive, a negative sign is multiplied with $\varrho.V.n.ds$ which means the mass is entering to R. We recall the divergence theorem once again to rewrite relation (4.4):

$$\int_V \text{div}.\vec{\text{V}}.\text{dV} = \int_S \text{n}.\vec{\text{V}}.\text{ds}$$

$$\iiint_V \frac{\partial \varrho}{\partial \text{t}} \text{dV} = -\iint_s \varrho \cdot \vec{V} \cdot \text{n} \cdot \text{ds} \equiv -\iiint_V \varrho \cdot \text{div} \cdot \vec{V} \cdot \text{dV}$$

$$\rightarrow \iiint_V \frac{\partial \varrho}{\partial t} dV + \iiint_V \varrho \cdot \text{div} \cdot \vec{V} \cdot dV = 0 \tag{4.5}$$

$$\rightarrow \iiint_V \left(\frac{\partial \varrho}{\partial \text{t}} + \varrho \text{div} \vec{\text{V}} \right) \text{dV} = \iiint_V \left(\frac{\partial \varrho}{\partial \text{t}} + \text{div}(\varrho \vec{\text{V}}) \right) \text{dV} = 0 \tag{4.6}$$

Relation (4.6) is satisfied when:

$$\rightarrow \frac{\partial \varrho}{\partial t} + \text{div}((\varrho \vec{\text{V}}) = 0 \tag{4.7}$$

Relation (4.7) is the continuity equation in Eulerian coordinates. We consider Eq. 4.8 once again:

$$\frac{\partial}{\partial t} + V_i . \frac{\partial}{\partial x_i} = \frac{D}{Dt} = 0 \tag{4.8}$$

Now, we have expanded relationships (4.7):

$$\frac{\partial \varrho}{\partial \text{t}} + \vec{\text{V}} \cdot \text{ dive } + \varrho \cdot \text{div } \vec{\text{V}} = 0$$

$$\rightarrow \frac{\partial \varrho}{\partial t} + V_i \cdot \frac{\partial \varrho}{\partial x_i} + \varrho \frac{\partial V_i}{\partial x_i} = 0 \tag{4.9}$$

Comparing relations (4.8) and (4.9), we conclude:

$$\frac{D\varrho}{Dt} + \varrho . \frac{\partial V_i}{\partial x_i} = 0 \tag{4.10}$$

Equation 4.10 represents the Eulerian form of the continuity equation. If the fluid is assumed incompressible, which means that its density is constant, then:

$$\frac{\text{D}\varrho}{\text{Dt}} = 0 \rightarrow \frac{\partial \text{ V}_\text{i}}{\partial \text{x}_\text{i}} = \text{div } \vec{\text{V}} = 0 \tag{4.11}$$

Relation (4.11) is the main condition for fluid to be incompressible.

The Material Time Derivative of a Volumetric Integral

Assume that φ is a physical quantity like mass or energy which depends on the particles of a continuum body. Now, the total amount of φ which is surrounded inside the environment R with volume V is:

$$\iiint_R \varphi \varrho dV \tag{4.12}$$

Relation (4.12) has been written for a time t and is calculated for this original time but as the time increases to $t + \delta t$ then amount of φ also varies inside R. This variation is shown using material time derivative as follows:

$$\frac{D}{Dt}\iiint_R \varphi \varrho dV \tag{4.13}$$

Relation (4.13) expresses the rate of variation of φ during the time according to number of the particles which pass through the environment R and its surface S. Referring to the mass conservation law the increase of mass rate of φ, inside R, is equal to summation of the existing mass particles, inside R, plus the input mass rate of φ that enters R:

$$\iiint_R \frac{\partial(\varphi R)}{\partial t} dV + \iint_s \varphi \varrho n.\vec{V} ds \equiv \frac{D}{Dt}\iiint_R \varphi \varrho dV \tag{4.14}$$

Using divergence theorem, we change the surface integral to volumetric integral for relation (4.14):

$$\frac{D}{Dt}\iiint_R \varphi \rho dV = \iiint_R \left\{\frac{\partial(\varphi \varrho)}{\partial t} + div\left(\varphi \varrho \vec{V}\right)\right\} dV \tag{4.15}$$

We recall Eulerian continuity equation and Eulerian derivative relation once again and rewrite relation (4.15). The right section of relation (4.15) is changed to $\varrho \frac{D\varphi}{Dt}$

Eulerian derivative relation: $\frac{D\varphi}{Dt} = \vec{V} \operatorname{grad} \varphi(x_i, t) + \frac{\partial \varphi(x_i, t)}{\partial t}$.

Eulerian continuity relation: $\frac{\partial \varrho}{\partial t} + \operatorname{div}\left(\varrho \vec{V}\right) = 0$

$$\rightarrow \varrho \frac{\partial \varphi}{\partial t} = \frac{\partial(\varphi \varrho)}{\partial t} + \operatorname{div}\left(\varphi \varrho \vec{V}\right) \equiv \varphi \left\{\frac{\partial \varrho}{\partial t} + \operatorname{div}\left(\varrho \vec{V}\right)\right\} + \varrho\left(\frac{\partial \varphi}{dt} + \vec{V} \cdot \operatorname{grad}\varphi\right) \tag{4.16}$$

Finally:

$$\frac{D}{Dt}\iiint_R \varphi \rho dV = \iiint_R \frac{D\varphi}{Dt} \varrho dV \tag{4.17}$$

Relation (4.17) is used in the next parts as an important relation.

Conservation of Linear Momentum

Based on the second law of Newton, the change of momentum of a particle with mass m is equal to the resultant external forces that are applied on it:

$$\sum \vec{\mathrm{F}} = \mathrm{m}.\vec{\mathrm{a}} \equiv \mathrm{m}\frac{\mathrm{dv}}{\mathrm{dt}} \equiv \frac{\mathrm{d(mv)}}{\mathrm{dt}} \tag{4.18}$$

In general, for a continuum environment R, the rate of change of linear momentum of the particles is proportional to all applied forces. The forces include the surface forces $t^{(n)}$ and volumetric ones b. Therefore, relation (4.18) is written as:

$$\frac{D}{Dt}\iiint_R \varrho \vec{V} dV = \iiint_R \varrho \vec{b} dV + \iint_s \vec{t}^{(n)} ds \tag{4.19}$$

Writing relation (4.19) based on index notation:

$$\frac{D}{Dt}\iiint_R \varrho V_i dV = \iiint_R \varrho b_i dV + \iint_s T_{ij}.n_i ds \tag{4.20}$$

Relation (4.17) is rewritten by putting vector velocity $\vec{V}$ instead of φ as φ can be replaced by any physical quantity:

$$\frac{D}{Dt}\iiint_R V_i \varrho dV \equiv \iiint_R \frac{DV_i}{Dt}\varrho dV \tag{4.21}$$

Comparing relations (4.18) and (4.19), we see:

$$\frac{D}{Dt}\iiint_R V_i \varrho dV = \iiint_R \frac{DV_i}{Dt}\varrho dV = \iiint_R \varrho b_i dV + \iint_s T_{ij}.n_i ds \tag{4.22}$$

Again, we use divergence theorem for relation (4.22) and change the surface integral to volumetric for: $\iint_s T_{ij}.n_i ds$

$$\iiint_\mathrm{R} \left\{\varrho\frac{\mathrm{DV_i}}{\mathrm{Dt}} - \varrho \mathrm{b_i} - \mathrm{div}\left(\mathrm{T_{ij}}\right)\right\}\mathrm{dV} = 0$$
$$\rightarrow \iiint_R \left\{\varrho\frac{DV_i}{Dt} - \varrho b_i - \frac{\partial T_{ij}}{\partial x_i}\right\}dV = 0$$
$$\rightarrow \frac{\partial T_{ij}}{\partial x_i} + \varrho b_i \equiv \varrho\frac{DV_i}{Dt} \tag{4.23}$$

Note that $\frac{DV_i}{Dt}$ is equal to concept of the acceleration vector a_i a_i

$$\rightarrow \frac{\partial T_{ij}}{\partial x_i} + \varrho b_i \equiv \varrho a_\mathrm{i} \tag{4.24}$$

Relation (4.24) is an important equation regarding motion of a continuum environment and if the acceleration vector is zero then relation (4.24) is changed to the equilibrium equations of a continuum environment that we have seen before in stress chapter.

$$\rightarrow \frac{\partial T_{ij}}{\partial x_i} + \varrho b_i \equiv 0 \tag{4.25}$$

Energy Conservation Law

The kinematic energy of a continuum environment which has occupied a space R is written according to below relation. The simple form of the relation is $\frac{1}{2}mv^2$ which is for a particle of mass m and the velocity magnitude of v but relation (4.26) is for the whole particles of the environment:

$$K = \frac{1}{2} \iiint_R \varrho V_i . V_i dV \tag{4.26}$$

Note that kinematic energy is part of the continuum environment energy and another part is the internal energy E, which is defined as:

$$E = \iiint_R \varrho e dV \tag{4.27}$$

where e is the energy per mass.

The energy conservation law says that the variations of kinematic and internal energy for a material that is surrounded in a space R is equal to the summation of mechanical work which is done by applied surface and volume forces as well as the work that is done by other entering internal energies (to R environment). These entering energies include the ones that come through the surface of the environment R like heat energy, radiation, electromagnetic, etc. If the heat energy vector that enters the environment R, as heating flux, is shown as q then q.n is the normal input heat flux (normal to the surface s) and n is the normal unit of s. Now, we can write the energy conservation law as below:

$$\frac{D}{Dt} \iiint_R \varrho (\frac{1}{2} V_i . V_i + e) dV = \iiint_R \varrho b_i V_i dV + \iint_S (T_{ji} . V_i - q_i) n_j d_s \tag{4.28}$$

Again, note that $q_{i.n_j d_s}$ has a negative value because the direction of q and n is not the same, q inward R and n is outward, therefore a negative sign shall be multiplied by $q_{i.n_j d_s}$ to let its value is positive. The positive value means that q enters the surface s. Now, relation (4.17) is applied for the left side of relation (4.28) and we use divergence theorem to change the surface integral to volume for right side of relation (4.28):

$$\frac{D}{Dt} \iiint_R \varrho (\frac{1}{2} V_i . V_i + e) dV = \iiint_R \frac{D}{Dt} \left\{ \varrho (\frac{1}{2} V_i . V_i + e) \right\} dV$$

$$\equiv \iiint_R (\varrho b_i V_i + div(T_{ji}.V_i - q_i)dV$$

$$\rightarrow \varrho \frac{D}{Dt}\left(\frac{1}{2}V_iV_i + e\right) = \varrho b_i V_i + diV(T_{ji}.V_i - q_i) \tag{4.29}$$

$$\rightarrow \varrho \frac{D}{Dt}\left(\frac{1}{2}V_iV_i + e\right) = \varrho b_i V_i + \frac{\partial}{\partial x_i}(T_{ji}.V_i - q_i) \tag{4.30}$$

The right side of relation (4.30) is differentiated where $\frac{DV_i}{Dt} = a_i$, then we rewrite the relation as:

$$\rightarrow V_i\left(\frac{\partial T_{ji}}{\partial x_j} + \varrho b_i - \varrho a_i\right) + \varrho \frac{De}{Dt} - T_{ji}.\frac{\partial V_i}{\partial x_j} + \frac{\partial q_i}{\partial x_j} = 0 \tag{4.31}$$

According to relation (4.24):

$$\frac{\partial T_{ij}}{\partial x_i} + \varrho b_i - \varrho a_i = 0$$

Therefore:

$$\rightarrow \varrho \frac{De}{Dt} - T_{ji}.\frac{\partial V_i}{\partial x_j} + \frac{\partial q_i}{\partial x_j} = 0$$

$$\rightarrow \varrho \frac{De}{Dt} = T_{ji}.\frac{\partial V_i}{\partial x_j} - \frac{\partial q_i}{\partial x_j} \tag{4.32}$$

Relation (4.32) expresses the energy equation for a continuum environment. We rewrite the relation as below:

$$T_{ji} \cdot \frac{\partial V_i}{\partial x_j} = T_{ij} \cdot \frac{\partial V_j}{\partial x_i} \quad : T_{ji} = T_{ij}$$

$$T_{ji}.\frac{\partial V_i}{\partial x_j} = \frac{1}{2}T_{ij}\left(\frac{\partial V_i}{\partial x_j} + \frac{\partial V_j}{\partial x_i}\right) \equiv T_{ij}.D_{ij}$$

$$\rightarrow \varrho \frac{De}{Dt} = T_{ij}.D_{ij} - \frac{\partial q_i}{\partial x_i} \tag{4.33}$$

D_{ij} is the tensor of the rate of deformation and the power of stress or rate of stress work is called $T_{ij}\,D_{ij}$.

According to Fourier's law, the heat flux can be written by:

$$q = -K.grad(T) \tag{4.34}$$

where T is the temperature and K is the heat conduction coefficient. Moreover, relation (4.33) is another expression of the first law of thermodynamics when the rate of heat transfer to environment R is zero.

Second Law of Thermodynamics

We first explain the concept of the second law of thermodynamics. Assume there is a cup of tea in cold weather where the tea loses its temperature after some time and gets cold. The first law of thermodynamics cannot give any information about the possibility of a spontaneous process of heating the tea, or whether the heat may be transferred from a cold source to a warm source spontaneously. Can such processes happen in reality? Therefore, the second law of thermodynamics helps to study the possibility of processes happening in nature.

Assume that the heat amount Q is produced by the environment R, the second law defines a property, namely entropy S, as below:

$$\varrho \frac{DS}{\mathrm{Dt}} \geq -\mathrm{div}\frac{\mathrm{q}}{\mathrm{T}} + \varrho \frac{\mathrm{Q}}{\mathrm{T}}$$

T is the absolute temperature, and Q is the input heat flow to R. In all thermodynamics processes, the main term and condition for a process to happen is that the summation of the system (or control volume) entropy and the environment entropy is equal to or greater than zero.

Problems

1. For the motion field $\begin{cases} x_1 = (1+at)x_1 \\ x_2 = X_2 \\ x_3 = X_3 \end{cases}$ with the density relation $\varrho = \frac{\varrho_0}{1+at}$

Which ϱ_0 is a constant value. Show that the continuity equation is satisfied.

$$\left.\begin{array}{r} x_1 = (1+at)x_1 \\ X_1 = \frac{x_1}{1+at} \end{array}\right\} \rightarrow V_1 = \overset{\circ}{x_1} = ax_1 \equiv a\left(\frac{x_1}{1+at}\right) = \frac{ax_1}{1+at}$$

$$x_2 = X_2 \rightarrow \overset{\circ}{x_2} = V_2 = 0$$

$$x_3 = X_3 \rightarrow \overset{\circ}{x_3} = V_3 = 0$$

The continuity equation in Eulerian coordinates:

$$\frac{D\varrho}{Dt} + \varrho div\left(\vec{V}\right) = 0 \rightarrow \frac{D}{Dt}\left(\frac{\varrho_0}{1+at}\right) + \varrho\left(\frac{\partial V_1}{\partial x_1} + \frac{\partial V_2}{\partial x_2} + \frac{\partial V_3}{\partial x_3}\right) = 0$$

$$\rightarrow \left\{ \begin{array}{c} \frac{-\varrho_0(a)}{(1+at)^2} + \varrho\left(\frac{a}{1+at}\right) = 0 \\ \frac{\partial V_1}{\partial x_1} = \frac{\partial}{\partial x_1}\left(\frac{ax_1}{1+at}\right) = \frac{a}{1+at} \\ \varrho = \left(\frac{\varrho_0}{1+at}\right) \end{array} \right\} \rightarrow$$

$$\rightarrow \frac{-\varrho_0 a}{(1+at)^2} + \left(\frac{\varrho_0}{1+at}\right)\left(\frac{a}{1+at}\right) = 0$$

$$\frac{-\varrho_0 a + \varrho_0 a}{(1+at)^2} = 0 \rightarrow 0 = 0$$

Therefore, the continuity equation is satisfied.

2. The stress tensor is given below for a body in equilibrium position where there is no volumetric force. Determine the values of b and c to satisfy the equilibrium equations.

$$T_{ij} = \begin{bmatrix} Cx_1^2 + bx_2^2 & 4x_1x_2 & 0 \\ 4x_1x_2 & -2bx_2 & 0 \\ 0 & 0 & 0 \end{bmatrix}$$

$$Equilibirium\,equations: \frac{\partial T_{ji}}{\partial x_j} + \varrho B_i = 0$$

$$\rightarrow \left\{ \begin{array}{l} \frac{\partial T_{11}}{\partial x_1} + \frac{\partial T_{12}}{\partial x_2} + \frac{\partial T_{13}}{\partial x_3} + \varrho B_1 = 0 \\ \rightarrow 2C\mathrm{x}_1 + 4\mathrm{x}_1 + 0 = 0 \rightarrow C = -2 \\ \frac{\partial T_{21}}{\partial x_1} + \frac{\partial T_{22}}{\partial x_2} + \frac{\partial T_{23}}{\partial x_3} + \varrho B_2 = 0 \\ 4\mathrm{x}_1 - 2b\mathrm{x}_2 = 0 \rightarrow b = 2 \end{array} \right.$$

3. The volumetric force vector is given $\vec{B} = -\mathrm{gi}_3$ as g is a constant value. Determine T_{33} in the stress tensor $T = \begin{bmatrix} x_1 & -x_3 & 0 \\ -x_3 & 0 & -x_2 \\ 0 & -x_2 & T_{33} \end{bmatrix}$ for satisfying the equilibrium equations.

$$\frac{\partial T_{ij}}{\partial x_j} + \varrho B_i = 0$$

$$\rightarrow^{i=3} \frac{\partial T_{31}}{\partial x_1} + \frac{\partial T_{32}}{\partial x_2} + \frac{\partial T_{33}}{\partial x_3} + \varrho B_3 = 0$$

$$\rightarrow 0 + (-1) + \frac{\partial T_{33}}{\partial x_3} + \varrho(-g) = 0$$

$$\frac{\partial T_{33}}{\partial x_3} = 1 + \varrho g \rightarrow T_{33} = (1 + \varrho g)x_3 + f(x_1, x_2)$$

The Theory of Linear Elasticity

5

The fundamental equations for linear elastic solids

In previous chapters, we discussed the concepts of stress and strain. In this chapter, a linear elastic material is investigated. The elastic material is a material that returns to its original form after being under stress and experiencing deformation. Linear elastic material is an ideal material that follows the Hook law. The Hook law is defined in its simple form as a relation between stress and strain for a single axial load:

$$\sigma = E.\varepsilon \tag{5.1}$$

The relation (5.1) is a linear relationship between stress and strain and any material which follows this relation is called a Hookean material. Our aim is to find the general form of the relation for three-dimensional situation for a continuum environment. In this regard, we use a internal energy function, u, for the volume of a continuum material V. In other words, we define the internal energy function per unit volume u_0 which is called the density of internal energy:

$$u = \int_v u_0 dv \tag{5.2}$$

The variation of the internal energy is calculated by:

$$\delta u = \int_v \delta u_0 dv \tag{5.3}$$

Noting to the first law of thermodynamics in absence of heat energy:

K. Hassani and F. Azimifar, *Introduction to Continuum Mechanics*, Synthesis Lectures on Mechanical Engineering, https://doi.org/10.1007/978-3-032-05076-2_5

$$\delta u = \delta w \tag{5.4}$$

$$\rightarrow \delta w = \int_v [\sigma_{11}.\delta e_{11} + \sigma_{22}.\delta e_{22} + \sigma_{33}.\delta e_{33} + 2\sigma_{12}.\delta e_{12} + 2\sigma_{13}.\delta e_{13} + 2\sigma_{23}.\delta e_{23}]dv \tag{5.5}$$

$$\rightarrow \delta u_0 = \sigma_{11}.\delta e_{11} + \sigma_{22}.\delta e_{22} + \sigma_{33}.\delta e_{33} + 2\sigma_{12}.\delta e_{12} + 2\sigma_{13}.\delta e_{13} + 2\sigma_{23}.\delta e_{23} \tag{5.6}$$

e indicates to the strain and relation (5.6) is written based on index notation:

$$\rightarrow \delta u_0 = \sigma_{ij}.\delta e_{ij} \tag{5.7}$$

In general, it can be said that the density of the strain energy is a function u_0 including nine components of the strain tensor:

$$u_0 = u_0(e_{11}, e_{22}, e_{33}, e_{12}, e_{13}, e_{21}, e_{23}, e_{31}, e_{32}) \tag{5.8}$$

The relation (5.8) in differential form is:

$$\begin{aligned} \delta u_0 = & \frac{\partial u_0}{\partial e_{11}}.\delta e_{11} + \frac{\partial u_0}{\partial e_{22}}.\delta e_{22} + \frac{\partial u_0}{\partial e_{33}}.\delta e_{33} + \frac{\partial u_0}{\partial e_{12}}.\delta e_{12} + \frac{\partial u_0}{\partial e_{13}}.\delta e_{13} \\ & + \frac{\partial u_0}{\partial e_{23}}.\delta e_{23} + \frac{\partial u_0}{\partial e_{21}}.\delta e_{21} + \frac{\partial u_0}{\partial e_{31}}.\delta e_{31} + \frac{\partial u_0}{\partial e_{32}}.\delta e_{32} \end{aligned} \tag{5.9}$$

Comparing relations (5.8) and (5.9):

$$\begin{aligned} \delta u_0 = & \sigma_{xx}.\delta e_{11} + \sigma_{22}.\delta e_{22} + \sigma_{33}.\delta e_{33} + 2\sigma_{12}.\delta e_{12} \\ & + 2\sigma_{13}.\delta e_{13} + 2\sigma_{y3}.\delta e_{23} \end{aligned} \tag{5.10}$$

$$\rightarrow \sigma_{ij} = \frac{\partial u_0}{\partial e_{ij}} \tag{5.11}$$

To show the stress–strain relationship in linear elastic bodies, we should define the function of density of strain energy as a polynomial of n order by e_{ij}:

$$u_\circ = C_\circ + C_{ij}.e_{ij} + C_{ijkl}e_{ij}e_{kl} + C_{ijklmn}\, e_{ij}e_{kl}e_{mn} + \ldots \tag{5.12}$$

On the other hand:

$$\rightarrow \sigma_{ij} = \frac{\partial u_0}{\partial e_{ij}} = C_{ij} + C_{ijkl}e_{kl} + C_{ijklmn}e_{kl}.e_{mn} + \ldots \tag{5.13}$$

It was confirmed to consider a linear relationship for the stress–strain of linear elastic bodies, therefore:

$$\rightarrow \sigma_{ij} = C_{ij} + C_{ijkl}.e_{kl} \tag{5.14}$$

In all elastic materials, the stress value is zero when all strain components are zero:

$$C_{ij} = 0 \rightarrow \sigma_{ij} = 0 \rightarrow C_{ij} = 0$$

Thus, relation (5.14) is defined as below:

$$\rightarrow \sigma_{ij} = C_{ijkl}.e_{kl} \tag{5.15}$$

C is a 9 × 9 tensor with 81 components but we will see later that the number of the non-identical components decreased to 21 due to some symmetrical properties in stress and strain tensors. The stress–strain tensor is shown as matrix form:

$$\begin{bmatrix} \sigma_{11} \\ \sigma_{22} \\ \sigma_{33} \\ \sigma_{12} \\ \sigma_{13} \\ \sigma_{23} \\ \sigma_{21} \\ \sigma_{31} \\ \sigma_{32} \end{bmatrix} = \begin{bmatrix} C_{1111} & C_{1122} & C_{1133} & C_{1112} & C_{1113} & C_{1123} & C_{1121} & C_{1131} & C_{1132} \\ C_{2211} & C_{2222} & C_{2233} & C_{2212} & C_{2213} & C_{2223} & C_{2221} & C_{2231} & C_{2232} \\ C_{1111} & C_{3322} & C_{3333} & C_{3312} & C_{3313} & C_{3323} & C_{3321} & C_{3331} & C_{3332} \\ C_{1211} & C_{1222} & C_{1233} & C_{1212} & C_{1213} & C_{1111} & C_{1221} & C_{1231} & C_{1232} \\ C_{1311} & C_{1322} & C_{1333} & C_{1312} & C_{1313} & C_{1223} & C_{1321} & C_{1331} & C_{1332} \\ C_{2311} & C_{2322} & C_{2333} & C_{2312} & C_{2313} & C_{2333} & C_{2321} & C_{2331} & C_{2332} \\ C_{2111} & C_{2122} & C_{2133} & C_{2112} & C_{2113} & C_{2123} & C_{2121} & C_{2131} & C_{2132} \\ C_{3111} & C_{3122} & C_{3133} & C_{3112} & C_{3113} & C_{3123} & C_{3121} & C_{3131} & C_{3132} \\ C_{3211} & C_{3222} & C_{3233} & C_{3212} & C_{3212} & C_{3223} & C_{3221} & C_{3231} & C_{3232} \end{bmatrix} \begin{bmatrix} e_{11} \\ e_{22} \\ e_{33} \\ e_{12} \\ e_{13} \\ e_{23} \\ e_{21} \\ e_{31} \\ e_{32} \end{bmatrix}$$

C is also called the matrix of elastic coefficient. We know that stress and strain tensors are both symmetrical, therefore:

$$\sigma_{kl} = \sigma_{lk} \rightarrow \begin{cases} \sigma_{kl} = C_{klmn}.e_{mn} \\ \sigma_{lk} = C_{lkmn}.e_{mn} \end{cases} \rightarrow C_{klmn} = C_{lkmn}$$

For example, $C_{1234} = C_{2134}$, and the components of C tensor are decreased to 54 due to this symmetric property. Now, we repeat the process for the strain tensor:

$$e_{mn} = e_{nm} \rightarrow \begin{cases} \sigma_{kl} = C_{klmn}.e_{mn} \\ \sigma_{kl} = C_{klnm}.e_{mn} \end{cases} \rightarrow C_{klmn} = C_{klnm}$$

Again, we see that some components of tensor C are identical, like $C_{1234} = C_{1243}$, due to symmetric property of strain tensor. Therefore 18 components of C are reduced and the total of non-identical components of C are 36 now.

$$C_{ijkl} = \begin{bmatrix} C_{1111} & C_{1122} & C_{1133} & C_{1112} & C_{1113} & C_{1123} \\ C_{2211} & C_{2222} & C_{2233} & C_{2212} & C_{2213} & C_{2223} \\ C_{3311} & C_{3322} & C_{3333} & C_{3312} & C_{3313} & C_{3323} \\ C_{1211} & C_{1222} & C_{1233} & C_{1212} & C_{1213} & C_{1223} \\ C_{1311} & C_{1322} & C_{1333} & C_{1312} & C_{1313} & C_{1323} \\ C_{2311} & C_{2322} & C_{2333} & C_{2312} & C_{2313} & C_{2323} \end{bmatrix} \tag{5.16}$$

We can show that current C is itself a symmetric tensor:

$$\begin{aligned} &\rightarrow \sigma_{ij} = \frac{\partial u_0}{\partial e_{ij}} = C_{ijkl}.e_{kl} \rightarrow \frac{\partial^2 u_0}{\partial e_{ij}.\partial e_{kl}} = C_{ijkl} \\ &\rightarrow \sigma_{kl} = \frac{\partial u_0}{\partial e_{kl}} = C_{klij}.e_{ij} \rightarrow \frac{\partial^2 u_0}{\partial e_{kl}.\partial e_{ij}} = C_{klij} \\ &\rightarrow \frac{\partial^2 \mathrm{u}_0}{\partial \mathrm{e}_{\mathrm{ij}}.\partial \mathrm{e}_{\mathrm{kl}}} = \frac{\partial^2 \mathrm{u}_0}{\partial \mathrm{e}_{\mathrm{kl}}.\partial \mathrm{e}_{\mathrm{ij}}} \rightarrow C_{ijkl} = C_{klij} \end{aligned} \tag{5.17}$$

The symmetry of C tensor reduces non- identical components are reduced to 21. In mathematics chapter, we introduced the transformation methods for calculating the components of a vector or tensor after rotation of the Cartesian coordinates in new one:

$$\acute{x}_k = n_{ik} \cdot x_i, \acute{\sigma}_{ikn} = n_{jk} \cdot n_{im} \cdot \sigma_{ij}$$

Similarly, the transformed components of a fourth order tensor, like C_{ijkl} can be obtained using the following relation:

$$C_{mnpq}\prime = n_{im} \cdot n_{jn} \cdot n_{kp} \cdot n_{lq} \cdot C_{ijkl} \tag{5.18}$$

We use relation (5.18) in the next sections to study the effects of symmetric planes and axes on C components.

The effects of the symmetric planes and axes on the elastic tensor C_{ijkl} components

Up to now, we found out that C tensor has 21 non-identical components, but many materials have some kind of symmetry in their internal structure which reduces the non-identical components of the tensor again. We explain below these materials as below:

1. Monotropic: A monoclinic material exhibits symmetric with respect to one plane, meaning its properties are identical on opposite sides of that plane.
2 Orthotropic: An orthotropic material exhibits symmetry about three mutually perpendicular planes, resulting in different material properties along each axis
3. Isotropic: An isotropic material has identical properties in all directions, while a transversely isotropic material exhibits symmetrical axis and is isotropic within the plane perpendicular to that axis.

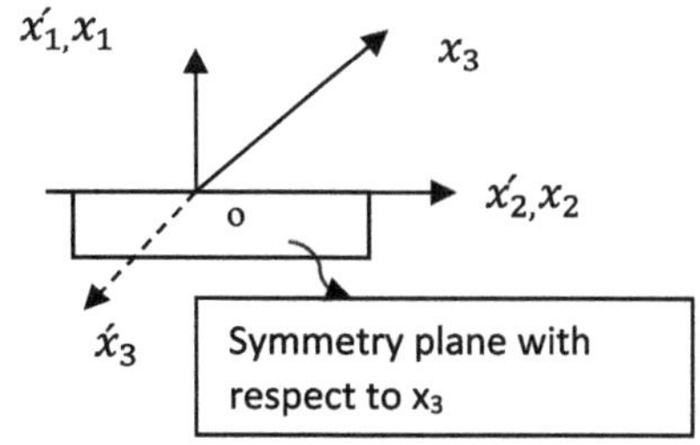

Fig. 5.1 Symmetry with respect to a plane for monotropic body

Monotropic bodies
Assume the body structure properties are symmetric with respect to x_3 axis. That means that there is no change in components of elastic tensor C with respect to the symmetry plane according to Fig. 5.1. As can be seen, the material properties of monotropic bodies are identical if coordinates axes are changed from $Ox_1x_2x_3$ to $O\acute{x}_1\acute{x}_2\acute{x}_3$

We recall the transformation tensor from the mathematics chapter which was used to transform the components of a vector or tensor from original coordinates system to rotated one: $a_{ij} = Cos\left(i_i', i_j\right)$.

Now, we intend to obtain a_{ij} for a monotropic material:

$$a_{ij} = \mathrm{Cos}\left(x_j', x_i\right) \rightarrow \begin{cases} a_{11} = \mathrm{Cos}\left(x_1', x_1\right) = \mathrm{Cos}0 = 1 \\ a_{12} = \mathrm{Cos}\left(x_1', x_2\right) = \mathrm{Cos}90^\circ = 0 \\ a_{13} = \mathrm{Cos}\left(x_1', x_3\right) = \mathrm{Cos}90^\circ = 0 \\ a_{21} = a_{12} = 0 \\ a_{31} = a_{13} = 0 \\ a_{22} = \mathrm{Cos}\left(x_2', x_2\right) = \mathrm{Cos}0 = 0 \\ a_{33} = \mathrm{Cos}\left(x_3', x_3\right) = \mathrm{Cos}180^\circ = -1 \\ a_{23} = \mathrm{Cos}\left(x_2', x_3\right) = \mathrm{Cos}90^\circ = 0 \\ = a_{32} \end{cases}$$

Therefore, for a monotropic body that its properties are identical with respect to x_3 according to Fig. 5.1 we write the transformation tensor a_{ij} as:

$$a_{ij} = \begin{bmatrix} 1 & 0 & 0 \\ 0 & 1 & 0 \\ 0 & 0 & -1 \end{bmatrix} \tag{5.19}$$

Now, we comeback to relation (5.18) and act based on a monotropic body where:

$$\acute{C}_{ijkm} = a_{ir} \cdot a_{js} \cdot a_{it} \cdot a_{mu} \cdot C_{rstu}$$

$$\begin{cases} \dot{C}_{1111} = a_{11} \cdot a_{11} \cdot a_{11} \cdot a_{11} \cdot C_{1111} = (1)(1)(1)(1)C_{1111} \equiv C_{1111} \\ \dot{C}_{2222} = a_{22} \cdot a_{22} \cdot a_{22} \cdot a_{22} \cdot C_{2222} = (1)(1)(1)(1)C_{2222} \equiv C_{2222} \\ \dot{C}_{3333} = a_{33} \cdot a_{33} \cdot a_{33} \cdot a_{33} \cdot C_{3333} = (-1)(-1)(-1)(-1) \equiv C_{3333} \\ \dot{C}_{1123} = a_{11} \cdot a_{11} \cdot a_{22} \cdot a_{33} \cdot C_{1123} = (1)(1)(1)(-1)C_{1123} \equiv -C_{1123} \end{cases}$$

As the components of C shall not be changed, all components of $\acute{C}_{ijkm}$ shall be the same as. C_{ijkl}. We see that $\acute{C}_{1123} = -C_{1123}$ but it should be $\acute{C}_{1123} = +C_{1123}$ to satisfy our condition for equality of all components of $\acute{C}_{ijkm}$ and C_{ijkl}. Therefore, the only justification is that C_{1123} value was zero originally. We have found out that the following components are also zero:

$$C_{1113} = C_{2213} = C_{2223} = C_{3313} = C_{3323} = C_{1213} = C_{1223} = 0$$

Finally, the monotropic elastic tensor has 13 non-identical components:

$$[C] = \begin{bmatrix} C_{1111} & C_{1122} & C_{1133} & C_{1112} & 0 & 0 \\ C_{1122} & C_{2222} & C_{2233} & C_{2212} & 0 & 0 \\ C_{1133} & C_{2233} & C_{3333} & C_{3312} & 0 & 0 \\ C_{1112} & C_{2212} & C_{3312} & C_{1212} & 0 & 0 \\ 0 & 0 & 0 & 0 & C_{1313} & C_{1323} \\ 0 & 0 & 0 & 0 & C_{1323} & C_{2323} \end{bmatrix} \quad (5.20)$$

We should emphasis again that when the coordinates system is transformed from $Ox_1x_2x_3$ to $O\acute{x}_1\acute{x}_2\acute{x}_3$ and there is no change in elastic properties with respect to x_3 axis, we mean that all components of C tensor which are in ox_3 are the same as C' in $o\acute{x}_3$ as per delimitation of a monotropic body. We had to show this physical issue using mathematics as above mentioned and showed in relation (5.20). If we repeat the obtained results with respect to the axes x_1 and x_2, then similar results are extracted. For example, if x_1 is chosen as the symmetry axis, then:

$$a_{ij} = \begin{bmatrix} -1 & 0 & 0 \\ 0 & 1 & 0 \\ 0 & 0 & -1 \end{bmatrix} \quad (5.21)$$

In this case, the elastic tensor C is as below with 13 non-identical components:

$$[C] = \begin{bmatrix} C_{1111} & C_{1122} & C_{1133} & C_{1123} & 0 & 0 \\ C_{1122} & C_{2222} & C_{2233} & C_{2223} & 0 & 0 \\ C_{1133} & C_{2233} & C_{3333} & C_{2333} & 0 & 0 \\ C_{1123} & C_{2223} & C_{2333} & C_{2323} & 0 & 0 \\ 0 & 0 & 0 & 0 & C_{1313} & C_{1213} \\ 0 & 0 & 0 & 0 & C_{1213} & C_{1212} \end{bmatrix} \quad (5.22)$$

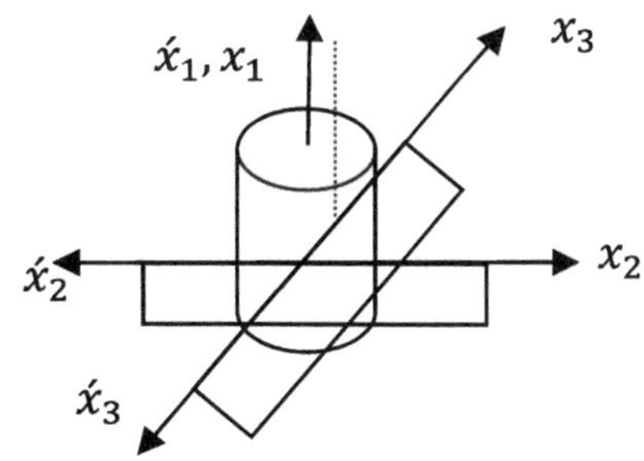

Fig. 5.2 Orthotropic Body

Orthotropic Bodies

An orthotropic body exhibits symmetrical about three mutually perpendicular planes, resulting in different material properties along each axis as shown in Fig. 5.2. The body has similar properties with respect to x_2 and x_3 axes and components of elastic tensor C are identical for the left and right sections of x_2 and x_3 axe.

We use the transformation tensor once again for an orthotropic body as below:

$$a_{ij} = \text{Cos}\left(\acute{x}_j, x_i\right)$$

$$Cos(\acute{x}_1, x_1) = Cos0^\circ = 1 \qquad Cos(\acute{x}_1, x_2) = Cos90^\circ = 0$$

$$Cos(\acute{x}_2, x_2) = Cos180^\circ = -1 \qquad Cos(\acute{x}_1, x_3) = Cos90^\circ = 0$$

$$Cos(\acute{x}_3, x_3) = Cos180^\circ = -1 \qquad Cos(\acute{x}_1, x_3) = Cos90^\circ = 0$$

$$Cos(\acute{x}_2, x_1) = Cos90^\circ = 0$$

$$a_{ij} = \begin{bmatrix} 1 & 0 & 0 \\ 0 & -1 & 0 \\ 0 & 0 & -1 \end{bmatrix} \tag{5.23}$$

$$\acute{C}_{ijkm} = a_{ir}.a_{js}.a_{kt}.a_{mu}.C_{rstu}$$

$$\acute{C}_{2323} = a_{22}.a_{33}.a_{22}.a_{33}.C_{2323} \equiv (-1)(-1)(-1)(-1)C_{2323} = C_{2323}$$

$$\acute{C}_{3131} = a_{33}.a_{11}.a_{33}.a_{11}.C_{3131} \equiv (-1)(1)(-1)(1)C_{3131} = C_{3131}$$

$$\acute{C}_{1111} = a_{11}.a_{11}.a_{11}.a_{11}.C_{1111} \equiv (1)(1)(1)(1)C_{1111} = C_{1111}$$

$$\acute{C}_{1112} = a_{11}.a_{11}.a_{11}.a_{22}.C_{1112} \equiv (1)(1)(1)(-1)C_{1112} = -C_{1112} \rightarrow C_{1112} = 0$$

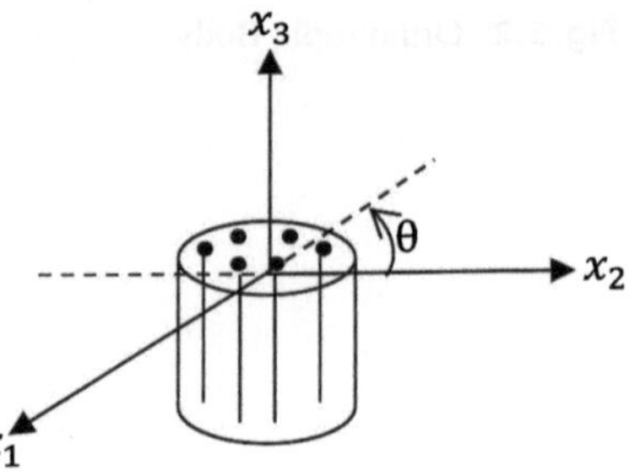

Fig. 5.3 The transversely isotropic body

Finally, there are only 9 non-identical components for the elastic tensor of an orthotropic body:

$$C_{ijkl} = \begin{bmatrix} C_{1111} & C_{1122} & C_{1133} & 0 & 0 & 0 \\ C_{1122} & C_{2222} & C_{2233} & 0 & 0 & 0 \\ C_{1133} & C_{2233} & C_{3333} & 0 & 0 & 0 \\ 0 & 0 & 0 & C_{1212} & 0 & 0 \\ 0 & 0 & 0 & 0 & C_{2323} & 0 \\ 0 & 0 & 0 & 0 & 0 & C_{3131} \end{bmatrix} \tag{5.24}$$

Similarly, if we choose the axes (x_1, x_3) and (x_1, x_2) as symmetry planes for an orthotropic body, we will reach the same results, and the elastic tensor C has 9 non-identical components. Wood is an example of an orthotropic material.

Transversely Isotropic Body

A transversely isotropic (also known as polar anisotropic) material is one with physical properties that are symmetric about an axis that is normal to a plane of isotropy. This transverse plane has infinite planes of symmetry and thus, within this plane, the material properties are the same in all directions. Figure 5.3 shows a transversely isotropic body.

We intend to express the concept of transversely isotropic bodies using mathematics. Note to Fig. 5.4 and the transformation tensor. Assume that the original coordinate system $x_1x_2x_3$ has rotated around the x_3 axis by θ. The definition of a transversely isotropic body says that the elastic tensor in new coordinates $\acute{x}_1\acute{x}_2\acute{x}_3$ shall be the same as in the original coordinates. Now we calculate the transformation tensor and elastic tensor for rotated coordinates $\acute{C}_{ijkl}$.

The transformation tensor is written based on Fig. 5.4:

$$a_{ij} = \mathrm{Cos}\left(\acute{x}_j, x_i\right)$$

$$a_{11} = \mathrm{Cos}(\acute{x}_1, x_1) = \mathrm{Cos}\theta$$

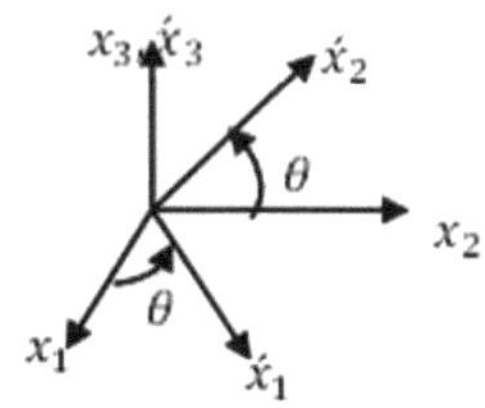

Fig. 5.4 The symmetry of the properties for a transversely isotropic material with respect to an arbitrary angle with respect to x3 axis

$$a_{12} = \mathrm{Cos}(\acute{x}_1, x_2) = \mathrm{Cos}\left(\frac{\pi}{2} - \theta\right) = \sin\theta$$

$$a_{13} = \mathrm{Cos}(\acute{x}_1, x_3) = \mathrm{Cos}\frac{\pi}{2} = 0$$

$$a_{21} = \mathrm{Cos}(\acute{x}_2, x_1) = \mathrm{Cos}\left(\frac{\pi}{2} + \theta\right) = -\sin\theta$$

$$a_{31} = \mathrm{Cos}(\acute{x}_3, x_1) = \mathrm{Cos}\left(\frac{\pi}{2}\right) = 0$$

$$a_{32} = \mathrm{Cos}(\acute{x}_3, x_2) = \mathrm{Cos}\left(\frac{\pi}{2}\right) = 0$$

$$a_{33} = \mathrm{Cos}(\acute{x}_3, x_3) = \mathrm{Cos}0 = 1$$

$$a_{ij} = \begin{bmatrix} Cos\theta & sin\theta & 0 \\ -sin\theta & Cos\theta & 0 \\ 0 & 0 & 1 \end{bmatrix} \tag{5.25}$$

The components of the relation (5.25) tensor are substituted in the elastic tensor of rotated coordinates:

$$\acute{C}_{ijkm} = a_{ir} \cdot a_{js} \cdot a_{kt} \cdot a_{mu} \cdot C_{rstu}$$

We calculate the following components for new elastic tensor:

$$\begin{aligned} \acute{C}_{1111} = & \left(\mathrm{Cos}^4\theta\right)C_{1111} + \left(\mathrm{Cos}^2\theta \sin^2\theta\right)(2E_{1122} + 4E_{1212}) \\ & + \left(\mathrm{Sin}^4\theta\right)C_{2222} \equiv C_{1111} \end{aligned} \tag{5.26}$$

$$\begin{aligned} \acute{C}_{1122} = & \left(Cos^2\theta sin^2\theta\right)C_{1111} + \left(Cos^4\theta\right)C_{1122} - 4\left(Cos^2\theta sin^2\theta\right)C_{1212} \\ & + \left(Sin^4\theta\right)C_{2211} + \left(sin^2\theta Cos^2\theta\right)C_{2222} \equiv C_{1122} \end{aligned} \tag{5.27}$$

$$\acute{C}_{1133} = \left(Cos^2\theta\right)C_{1133} + \left(sin^2\theta\right)C_{2233} \equiv C_{1133} \tag{5.28}$$

$$\acute{C}_{2222} = \left(sin^4\theta\right)C_{1111} + \left(sin^2\theta Cos^2\theta\right)(2C_{1122} + 4C_{1212}) + \left(Cos^4\theta\right)C_{2222} \equiv C_{2222} \tag{5.29}$$

$$\begin{aligned}\acute{C}_{1212} = &\left(Cos^2\theta sin^2\theta\right)C_{1111} - 2\left(Cos^2\theta sin^2\theta\right)C_{1122} - 2\left(Cos^2\theta sin^2\theta\right)C_{1212} \\ &+ \left(Cos^4\theta\right)C_{1212} + \left(sin^2\theta Cos^2\theta\right)C_{2222} + \left(sin^4\theta\right)C_{1212} \equiv C_{1212}\end{aligned} \tag{5.30}$$

The components of the original elastic tensor shall be the same as the new elastic tensor for a transversely isotropic condition, for example: $\backslash acuteC_{1122} = C_{1111}$.

Solving Eqs. 5.26 to 5.30 gives interesting results. for instance, we solve Eq. 5.28:

$$\acute{C}_{1133} = \left(Cos^2\theta\right)C_{1133} + \left(sin^2\theta\right)C_{2233} \equiv C_{1133}$$

The equation is only valid when $C_{2233} = C_{1133}$:

$$C_{1133} = \left(Cos^2\theta\right)C_{1133} + \left(sin^2\theta\right)C_{1133} \equiv \left(sin^2\theta + Cos^2\theta\right)C_{1133}$$

$$\rightarrow C_{1133} = C_{1133}$$

The result of Eq. 5.28 is that $C_{1133} = C_{2233}$. By solving other equations:

$$\begin{cases} C_{1111} = C_{2222} \\ C_{1212} = \frac{1}{2}(C_{1111} - C_{1122}) \end{cases} \tag{5.31}$$

Similarly, we can write for other components like $\acute{C}_{1323}$

$$\begin{cases} \acute{C}_{1323} = C_{1323} = 0 \\ \acute{C}_{1323} = a_{11}.a_{12}.C_{1313} + a_{21}.a_{22}.C_{2323} \\ \equiv Cos\theta sin\theta C_{1313} + (-sin\theta)(Cos\theta)C_{2323} \\ \equiv Cos\theta sin\theta(C_{1313} - C_{2323}) \equiv 0 \\ \rightarrow C_{1313} = C_{2323} \end{cases} \tag{5.32}$$

Therefore, a transversely isotropic tensor has only 5 non-identical components:

$$C_{ijkl} = \begin{bmatrix} C_{1111} & C_{1122} & C_{1133} & 0 & 0 & 0 \\ C_{1122} & C_{1111} & C_{1133} & 0 & 0 & 0 \\ C_{1133} & C_{1133} & C_{3333} & 0 & 0 & 0 \\ 0 & 0 & 0 & C_{1313} & 0 & 0 \\ 0 & 0 & 0 & 0 & C_{1313} & 0 \\ 0 & 0 & 0 & 0 & 0 & \frac{1}{2}(C_{1111} - C_{1122}) \end{bmatrix} \tag{5.33}$$

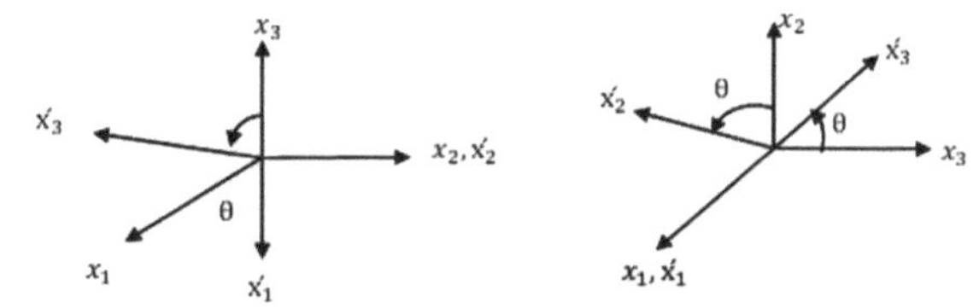

Fig. 5.5 Obtaining the elastic tensors for an isotropic material using two rotated coordinates with an angle θ

Isotropic Body

An isotropic body has identical properties in all directions. We use Fig. 5.5 to obtain the elastic tensor C for an isotropic body. Therefore, we consider two new rotated coordinates around x_1 and x_2 by an angle θ, and then obtain the transformation tensors as:

The transformation tensor for rotated coordinate systems around x_2:

$$a_{ij} = \begin{bmatrix} Cos\theta & 0 & -sin\theta \\ 0 & 1 & 0 \\ sin\theta & 0 & Cos\theta \end{bmatrix} \tag{5.34}$$

The transformation tensor for rotated coordinate systems around x_1:

$$a_{ij} = \begin{bmatrix} 1 & 0 & 0 \\ 0 & Cos\theta & sin\theta \\ 0 & -sin\theta & Cos\theta \end{bmatrix} \tag{5.35}$$

We substitute the above transformation tensor components in the following relation:

$$\acute{C}_{ijkm} = a_{ir} \cdot a_{js} \cdot a_{kt} \cdot a_{mu} \cdot C_{rstu}$$

$$a_{ij} = \begin{bmatrix} Cos\theta & 0 & sin\theta \\ 0 & 1 & 0 \\ sin\theta & 0 & Cos\theta \end{bmatrix} \rightarrow \begin{cases} C_{1111} = C_{3333} \\ C_{3131} = \frac{1}{2}(C_{1111} - C_{1133}) \end{cases} \tag{5.36}$$

$$a_{ij} = \begin{bmatrix} 1 & 0 & 0 \\ 0 & Cos\theta & sin\theta \\ 0 & -sin\theta & Cos\theta \end{bmatrix} \rightarrow \begin{cases} C_{1122} = C_{1133} \\ C_{2323} = \frac{1}{2}(C_{2222} - C_{2233}) \\ C_{1133} = C_{2233} \\ C_{3131} = \frac{1}{2}(C_{3333} - C_{1133}) \end{cases} \tag{5.37}$$

Similarly, we have:

$$\begin{cases} C_{1313} = \frac{1}{2}(C_{1111} - C_{1122}) \\ C_{3333} = C_{1111} \\ C_{1133} = C_{1122} \end{cases} \tag{5.38}$$

The elastic tensor of an isotropic body has only two non-identical components:

$$C_{ijkl} = \begin{bmatrix} C_{1111} & C_{1122} & C_{1122} & 0 & 0 & 0 \\ C_{1122} & C_{1111} & C_{1122} & 0 & 0 & 0 \\ C_{1122} & C_{1122} & C_{1111} & 0 & 0 & 0 \\ 0 & 0 & 0 & \frac{C_{1111}-C_{1122}}{2} & 0 & 0 \\ 0 & 0 & 0 & 0 & \frac{C_{1111}-C_{1122}}{2} & 0 \\ 0 & 0 & 0 & 0 & 0 & \frac{C_{1111}-C_{1122}}{2} \end{bmatrix} \quad (5.39)$$

Another Presentation of Elastic Tensor Components for an Isotropic Body
Lame introduced two parameters for the elastic tensor of an isotropic body, which are called Lame coefficients and are shown by λ and $\mu\mu$.

$$\begin{cases} C_{1122} = \lambda \\ \frac{1}{2}(C_{1111} - C_{1122}) = \mu \end{cases} \rightarrow \mathrm{C}_{1111} = 2\mu + \lambda \quad (5.40)$$

Now, we put these coefficients in the tensor of relation (5.39) instead of C_{1111} and C_{1122}:

$$C_{ijkl} = \begin{bmatrix} \lambda + 2\lambda & \lambda & \lambda & 0 & 0 & 0 \\ \lambda & \lambda + 2\mu & \lambda & 0 & 0 & 0 \\ \lambda & \lambda & \lambda + 2\mu & 0 & 0 & 0 \\ 0 & 0 & 0 & 2\mu & 0 & 0 \\ 0 & 0 & 0 & 0 & 2\mu & 0 \\ 0 & 0 & 0 & 0 & 0 & 2\mu \end{bmatrix} \quad (5.41)$$

This tensor can be written by index notation relation as below:

$$C_{ijkl} = \lambda\delta_{ij}.\delta_{km} + \mu\left(\delta_{ik}.\delta_{jm} + \delta_{im}.\delta_{kj}\right) \quad (5.42)$$

For example:

$$\begin{aligned} C_{1111} &= \lambda\delta_{11}.\delta_{11} + \mu(\delta_{11}.\delta_{11} + \delta_{11}.\delta_{11}) \\ &= \lambda(1)(1) + \mu(1 \times 1 + 1 \times 1) \\ &= \lambda + 2\mu \end{aligned}$$

Or:

$$\begin{aligned} C_{1122} &= \lambda\delta_{11}.\delta_{22} + \mu(\delta_{12}.\delta_{12} + \delta_{12}.\delta_{21}) \\ &= \lambda(1)(1) + \mu(0 + 0) = \lambda \end{aligned}$$

It is the time to express the relation between stress and strain based on Hook law:

$$[\sigma] = [C].[\mathrm{e}]$$

We put relation (5.42) instead of C and write relation (5.43) based on index notation:

$$
\begin{aligned}
\sigma_{ij} &= \left[\lambda\delta_{ij}.\delta_{km} + \mu\left(\delta_{ik}.\delta_{jm} + \delta_{im}.\delta_{jk}\right)\right]e_{\mathrm{km}} \\
&= \left[\lambda\delta_{ij}.\delta_{km}.e_{km} + \mu\left(\delta_{ik}.\delta_{jm}.e_{km} + \delta_{im}.\delta_{jk}.e_{\mathrm{km}}\right)\right] \\
&= \lambda\delta_{ij}.e_{kk} + \mu\left(e_{ij} + e_{\mathrm{ij}}\right) \\
&\rightarrow \sigma_{ij} = \lambda\delta_{ij}.e_{kk} + 2\mu e_{ij}
\end{aligned}
\tag{5.43}
$$

Relation (5.43) shows the relation between stress and strain for an isotropic material. By manipulating relation (5.43), we calculate stress:

$$
\sigma_{ii} = \lambda\delta_{ii}.e_{kk} + 2\mu e_{ii} = 3\lambda e_{kk} + 2\mu e_{ii} = (3\lambda + 2\mu)e_{ii} \tag{5.44}
$$

Note that e_{kk} and e_{ii} are similar and equal to $e_{11} + e_{22} + e_{33}$
Now, we substitute relation (5.44) into (5.43) and obtain the strain:

$$
e_{ij} = \frac{\sigma_{ij}}{2\mu} - \frac{\lambda\delta_{ij}}{2\mu(3\lambda + 2\mu)}.\sigma_{kk} \tag{5.45}
$$

Relation (5.45) shows again the stress–strain relation for an isotropic body. Now, we aim to present how Lame coefficients are calculated by Young module E and Poisson's coefficient ν.

We consider a rod which is only under an axial load in the direction of x_1, therefore:

$$
\sigma_{11} = \sigma,\ \sigma_{12} = \sigma_{11} = \sigma_{13} = \sigma_{33} = \sigma_{22} = 0
$$

The value of strain is calculated by relation (5.45):

$$
\begin{aligned}
e_{11} &= \frac{\lambda + \mu}{\mu(3\lambda + 2\mu)}\sigma \\
e_{22} = e_{33} &= \frac{-\lambda}{2\mu(3\lambda + 2\mu)}.\sigma \\
e_{12} = e_{23} &= e_{13} = 0
\end{aligned}
\tag{5.46}
$$

We know from strength of material science that E or module of elasticity, Young module, and Poisson's coefficient are calculated by:

$$
\begin{aligned}
\frac{\sigma}{e} &= E(Youngmodule) \\
\nu &= \frac{-e_{22}}{e_{11}} = \frac{-e_{33}}{e_{11}}
\end{aligned}
\tag{5.47}
$$

Comparing relations (5.46) and (5.47);

$$
\frac{1}{E} = \frac{\lambda + \mu}{\mu(3\lambda + 2\mu)},\ \upsilon = \frac{\lambda}{2(\lambda + \mu)} \tag{5.48}
$$

If we extract Lame coefficients form relation (5.48):

$$\begin{cases} \mu = \frac{E}{2(1+\nu)} \\ \lambda = \frac{\upsilon E}{(1+\nu)(1-2\nu)} \end{cases} \tag{5.49}$$

Relation (5.49) can be substituted into relation (5.45) and we obtain:

$$e_{ij} = \frac{-\nu}{E}.\sigma_{kk}.\delta_{ij} + \frac{1+\nu}{E}.\sigma_{ij} \tag{5.50}$$

We stress once again that relation (5.50) is applicable only for an isotropic material.

In the next step, we consider a body which is only under pure torsion $\tau = \sigma_{21} = \sigma_{12}$ and again from relation (5.45):

$$e_{12} = \frac{(1+\nu)\tau}{E} = \frac{\tau}{2\mu} \tag{5.51}$$

In this application, μ is called module of shear or G and:

$$G = \frac{E}{2(1+\nu)}$$

In third step, we assume a body is under a uniform stress condition like hydrostatic stress (P), therefore we have:

$$\sigma_{ij} = -P\delta_{ij} = \begin{bmatrix} -P & 0 & 0 \\ 0 & -P & 0 \\ 0 & 0 & -P \end{bmatrix} \tag{5.52}$$

By applying relation (5.52) on (5.45):

$$\begin{aligned} e_{ij} &= \frac{\sigma_{ij}}{3\lambda+2\mu} = \frac{-3P}{3\lambda+2\mu} = \frac{-P}{K} \\ \rightarrow K &= \frac{3\lambda+2\mu}{3} = \lambda + \frac{2\mu}{3} \end{aligned} \tag{5.53}$$

It should be noted that K is called Bulk modules and can be presented also as:

$$\begin{cases} K = \lambda + \frac{2\mu}{3} \\ \lambda = \frac{\nu E}{(1+\nu)(1-2\nu)} \\ \mu = \frac{E}{2(1+\nu)} \end{cases} \rightarrow K = \frac{E}{3(1-2\nu)} \tag{5.54}$$

Palin Stress and Strain

In many engineering applications, three-dimensional stress–strain problems can be studied as two dimensional or plain ones. For example, if one of the dimensions is very small compared to others then the stress can be neglected in the smallest dimension and this condition is called plain stress. On the other hand, if one of the dimensions is very big compared to others, then it is possible to neglect the strain in the biggest dimension and the condition is called plain strain.

In plain stress, we assume that there is no stress in x_3 direction for an isotropic body:

$$\sigma_{33} = \sigma_{31} = \sigma_{32} = 0$$

$$\sigma_{ij} = \lambda e_{kk}.\delta_{ij} + 2\mu e_{ij}$$

$$\begin{aligned}\sigma_{33} = 0 &= \lambda(e_{11} + e_{22} + e_{33})\delta_{33} + 2\mu e_{33} \\ &= \lambda(e_{11} + e_{22} + e_{33}) + 2\mu e_{33}\end{aligned}$$

$$\rightarrow e_{33} = \left(\frac{-\lambda}{\lambda + 2\mu}\right)(e_{11} + e_{22}) \tag{5.55}$$

Using relation (5.55) and putting σ_{12} and σ_{22} in relation (5.43), substituting λ and μ in terms of E and ν):

$$\begin{bmatrix}\sigma_{11}\\ \sigma_{22}\\ \sigma_{12}\end{bmatrix} = \frac{E}{1-\nu^2}\begin{bmatrix}1 & \nu & 0\\ \nu & 1 & 0\\ 0 & 0 & \frac{1-\nu}{2}\end{bmatrix}\begin{bmatrix}e_{11}\\ e_{22}\\ 2e_{12}\end{bmatrix} \tag{5.56}$$

Using a similar method for plain strain, $e_{33} = 0$:

$$\begin{bmatrix}\sigma_{11}\\ \sigma_{22}\\ \sigma_{12}\end{bmatrix} = \frac{E(1-\nu)}{(1-2\nu)(1+\nu)}\begin{bmatrix}1 & \frac{\nu}{1-\nu} & 0\\ \frac{\nu}{(1-\nu)} & 1 & 0\\ 0 & 0 & \frac{1-2\nu}{2(1-\nu)}\end{bmatrix}\begin{bmatrix}e_{11}\\ e_{22}\\ 2e_{12}\end{bmatrix} \tag{5.57}$$

In the end of chapter, it would be interesting to introduce two important relations one for stress in Newtonian fluids and Navier–stokes equation. You can refer to different fluid mechanics books to see how these relations were extracted.

The stress tensor in the fluids based on thermodynamics pressure of P is calculated by:

$$\sigma_{ij} = -\left[P + \left[\frac{2}{3}\mu - \lambda\right)\nabla.\vec{V}\right]\delta_{ij} + \mu\left(\frac{\partial v_i}{\partial x_j} + \frac{\partial v_j}{\partial x_i}\right) \tag{5.58}$$

Note that $\nabla.\vec{V}$ is for the dilation of fluids particles and expresses the rate of volume variations for fluid particles.

Navier-stokes equation is very important in fluid mechanics and is obtained by applying the Newtonian second law for motion of the fluid. This equation says that the induced stress on the fluid is due to viscosity and pressure of the fluid:

$$\varrho.\frac{DV_i}{Dt} = -\frac{\partial}{\partial x_i}\left[P + \left(\frac{2}{3}\mu - \lambda\right)\nabla.\vec{V}\right] + \frac{\partial}{\partial x_j}\left[\mu\left(\frac{\partial V_j}{\partial x_i} + \frac{\partial V_i}{\partial x_j}\right)\right] + \varrho G_i \qquad (5.59)$$

In this relation, G_i is the external volume force which is exerted per unit of mass.

Problems

For given displacement field, calculate the related components of stress tensor. Assume an isotropic material with very small strains.

$$u_1 = \mathrm{KX_2X_3}$$

$$u_2 = \mathrm{KX_1X_3}$$

$$u_3 = \mathrm{KX_1X_2}$$

As the material is isotropic, we use $\sigma_{ij} = \lambda\delta_{ij} + e_{kk} + 2\mu e_{ij}$ and we also use Green–Lagrange tensor for strain calculations, but the non-linear part of the tensor is zero because of very small strains:

$$e_{ij} = \frac{1}{2}\left(\frac{\partial u_i}{\partial x_j} + \frac{\partial u_j}{\partial x_i}\right)$$

$$e_{11} = \frac{\partial u_1}{\partial x_1} = 0,\ e_{22} = \frac{1}{2} \times 2 \times \frac{\partial u_2}{\partial x_2} = 0$$

$$e_{33} = \frac{\partial u_3}{\partial x_3} = 0$$

$$e_{12} = e_{21} = \frac{1}{2}\left(\frac{\partial u_1}{\partial x_1} + \frac{\partial u_2}{\partial x_1}\right) = \frac{1}{2}(\mathrm{KX_3} + \mathrm{KX_3}) = \mathrm{KX_3}$$

$$e_{13} = e_{31} = \frac{1}{2}\left(\frac{\partial u_1}{\partial x_3} + \frac{\partial u_3}{\partial x_1}\right) = \frac{1}{2}(\mathrm{KX_2} + \mathrm{KX_2}) = \mathrm{KX_2}$$

$$e_{23} = e_{32} = \frac{1}{2}\left(\frac{\partial u_2}{\partial x_3} + \frac{\partial u_3}{\partial x_2}\right) = \frac{1}{2}(\mathrm{KX_1} + \mathrm{KX_1}) = \mathrm{KX_1}$$

$$\begin{aligned}\sigma_{ij} &= \lambda\delta_{ij}.e_{kk} + 2\mu e_{ij} = \lambda.\delta_{ij}(e_{11} + e_{22} + e_{33}) + 2\mu e_{ij}\\ &= \lambda.\delta_{ij}(0 + 0 + 0) + 2\mu e_{ij} = 2\mu e_{ij}\end{aligned}$$

$$\begin{cases}\sigma_{12} = 2\mu e_{12} = 2\mu KX_3\\ \sigma_{13} = 2\mu e_{13} = 2\mu KX_2\\ \sigma_{23} = 2\mu e_{23} = 2\mu KX_1\\ \sigma_{11} = \sigma_{22} = \sigma_{33} = 0\end{cases} \rightarrow \sigma_{ij} = 2\mu K\begin{bmatrix}0 & X_3 & X_2\\ X_3 & 0 & X_1\\ X_2 & X_1 & 0\end{bmatrix}$$

2. An isotropic cube with side of 1 m is in water at a depth of 500 m. if Young module of the cube metal is E = 1000 Kpa and Poisson's ratio is 0.1, find the volume variation if the cube has no motion.

$$\varrho = 1000\, kg/m^3,\, g = 10\, m/s^2$$

When the cube has no motion, then the only applied stress is hydrostatic, therefore:

$$P = \frac{F}{A} = \varrho gh = 1000 \times 10 \times 500 = 5 \times 10^6 Pa = 5Mpa$$

$$\sigma_{ij} = \begin{bmatrix} -5 & 0 & 0 \\ 0 & -5 & 0 \\ 0 & 0 & -5 \end{bmatrix} Mpa$$

$$\varepsilon_x = \varepsilon_y = \varepsilon_z = \frac{1}{E}\left[\sigma_x - \nu\left(\sigma_y + \sigma_z\right)\right]$$

This relation was extracted directly from relation (5.50). Now, we have a hydroelectric pressure condition $\sigma_{11} = \sigma_{22} = \sigma_{33}$ and $\sigma_{12} = \sigma_{13} = \sigma_{23} = 0$:

$$\begin{aligned} \varepsilon_{11} = \varepsilon_{22} = \varepsilon_{33} &= \frac{1}{E}[\sigma_{11} - \nu(\sigma_{22} + \sigma_{33})] \\ &\equiv \frac{1}{1000^{\text{kpa}}}\left[-5000^{\text{kpa}} - 0.1\left(-5000^{\text{kpa}} - 5000^{\text{kpa}}\right)\right] \\ &= \frac{-5000 + 1000}{1000} = -4 \end{aligned}$$

$$\rightarrow \frac{\Delta \text{V}}{\text{V}} = \varepsilon_{11} + \varepsilon_{22} + \varepsilon_{33} = -4 - 4 - 4 = -12$$

$$\rightarrow \Delta V = (-12)(1 \times 1 \times 1) = -12m^3$$

3. An isotropic plate is under stress as per shown. If there is no displacement and vorticity in the origin of the coordinate system at (0,0,0) then find the displacement functions. Assume the strains are small.

$$\begin{cases} \sigma_{11} = \sigma \\ \sigma_{22} = \sigma \\ \sigma_{12} = \sigma_{21} = \tau \end{cases}$$

$$\text{Isotropic body} \rightarrow e_{ij} = \frac{-\nu}{E}\sigma_{kk}.\delta_{ij} + \frac{1+\nu}{E}.\sigma_{ij}$$

$$e_{11} = \frac{-\nu}{E}(\sigma_{11} + \sigma_{22} + \sigma_{33}).\delta_{11} + \frac{1+\nu}{E}.\sigma_{11}$$

$$= \frac{-\nu}{E}\sigma + \sigma + 0) + \frac{1+\nu}{E}.\sigma = \frac{\sigma(1-\nu)}{E}$$

$$e_{22} = \frac{-\nu}{E}(\sigma + \sigma).\delta_{22} + \frac{1+\nu}{E}.\sigma = \frac{\sigma(1-\nu)}{E}$$

$$e_{33} = \frac{-\nu}{E}(\sigma + \sigma).\delta_{33} + \frac{1+\nu}{E}.(0) = \frac{-2\sigma\nu}{E}$$

$$e_{13} = e_{23} = 0$$

$$e_{12} = \frac{-\nu}{E}(\sigma + \sigma).(0) + \frac{1+\nu}{E}.\sigma_{12} = \left(\frac{1+\nu}{E}\right)\tau$$

$$\text{small strains : } u_{i,k}.\mathrm{u}_{j,k} = 0$$

$$\rightarrow e_{ij} = \frac{1}{2}\left(\frac{\partial u_i}{\partial X_j} + \frac{\partial u_j}{\partial X_i}\right)$$

$$e_{11} = \frac{1}{2}\left(\frac{\partial u_1}{\partial X_1} + \frac{\partial u_2}{\partial X_1}\right) = \frac{\partial u_1}{\partial X_1} = \frac{\sigma(1-\nu)}{E}$$

$$\rightarrow u_1 = \int \frac{\sigma(1-\nu)}{E}\mathrm{dX}_1 = \frac{\sigma(1-\nu)}{E}X_1 + f_1(X_2)$$

$$e_{22} = \frac{1}{2}\left(\frac{\partial u_2}{\partial X_2} + \frac{\partial u_2}{\partial X_2}\right) = \frac{\partial u_2}{\partial X_2} \rightarrow \frac{\partial u_2}{\partial X_2} = \frac{\sigma(1-\nu)}{E} \rightarrow$$

$$\rightarrow u_1 = \int \frac{\sigma(1-\nu)}{E}\mathrm{dX}_2 = \frac{\sigma(1-\nu)}{E}X_2 + f_2(X_1)$$

$$\rightarrow u_2 = \int \frac{\sigma(1-\nu)}{E}\mathrm{dX}_2 = \frac{\sigma(1-\nu)}{E}X_2 + f_2(X_1)$$

$$e_{12} = \frac{1}{2}\left(\frac{\partial u_1}{\partial X_2} + \frac{\partial u_2}{\partial X_1}\right) = 2\frac{\sigma(1+\nu)}{E}\tau = \frac{\partial u_1}{\partial X_1} + \frac{\partial u_2}{\partial X_2}$$

$$2\frac{(1+\nu)}{E}\tau = \acute{f}_1(\mathrm{X}_2) + \acute{f}_2(\mathrm{X}_2)$$

$f_2(\mathrm{X}_2), f_1(\mathrm{X}_2)a$ re first order polynomials.

$$\begin{matrix} f_1(\mathrm{X}_2) = A_2X_2 + B_2 \\ f_2(\mathrm{X}_1) = A_1X_1 + B_1 \end{matrix} \rightarrow 2\frac{(1+\nu)}{E}\tau = A_2 + \mathrm{A}_1$$

$$\rightarrow u_1(0,0) = u_1(0,0) = 0 \rightarrow \begin{cases} u_1 = \frac{\sigma(1-\nu)}{E}(0) + f_1(X_2) = 0 \\ u_2 = \frac{\sigma(1-\nu)}{E}(0) + f_2(X_1) = 0 \end{cases}$$

$$\rightarrow \begin{cases} 0 = 0 + A_2(0) + B_2 \\ 0 = 0 + A_1(0) + B_1 \end{cases} \rightarrow B_1 = B_2 = 0$$

$$\rightarrow w(0,0) = \frac{\partial u_1}{\partial X_2} - \frac{\partial u_2}{\partial X_1} = 0 \rightarrow A_1 = A_2$$

$$A_1 = A_2 = \left(\frac{1+\nu}{E}\right)\tau \rightarrow$$

$$\rightarrow \begin{cases} u_1 = \frac{\sigma(1-\nu)}{E}X_1 + \left(\frac{1+\nu}{E}\right)\tau X_2 \\ u_1 = \frac{\sigma(1-\nu)}{E}X_2 + \left(\frac{1+\nu}{E}\right)\tau X_1 \end{cases}$$

4. An isotropic sphere with radius of 5 cm is under stress field as below. Calculate the volume change of the sphere.

$$T_{ij} = \begin{bmatrix} 6 & 2 & 0 \\ 2 & -3 & 0 \\ 0 & 0 & 0 \end{bmatrix}$$

$$\nu = 0.3; \quad \mu = 79.2Gpa; \quad E = 207Gpa$$

$$e_{ij} = \frac{-\nu}{E}(\sigma_{kk}).\delta_{ij} + \frac{(1+\nu)}{E}.\sigma_{ij}$$

$$\begin{aligned} e_{11} &= \frac{-\nu}{E}(\sigma_{11} + \sigma_{22} + \sigma_{33}).\delta_{11} + \frac{(1+\nu)}{E}.\sigma_{11} \\ &= \frac{-0.3}{207 \times 10^3}(6 - 3 + 0)(1) + \frac{1+0.3}{207 \times 10^3}(6) = 0.033 \times 10^{-3} \end{aligned}$$

$$\begin{aligned} e_{22} &= \frac{-\nu}{E}(\sigma_{11} + \sigma_{22} + \sigma_{33}).\delta_{22} + \frac{(1+\nu)}{E}.\sigma_{22} \\ &= \frac{-0.3}{207 \times 10^3}(6 - 3 + 0)(1) + \frac{1+0.3}{207 \times 10^3}(-3) = -0.0232 \times 10^{-3} \end{aligned}$$

$$\begin{aligned} e_{33} &= \frac{-\nu}{E}(\sigma_{11} + \sigma_{22} + \sigma_{33}).\delta_{33} + \frac{(1+\nu)}{E}.\sigma_{33} \\ &= \frac{-0.3}{207 \times 10^3}(6 - 3 + 0)(1) + \frac{1+0.3}{207 \times 10^3}(0) = -0.0126 \times 10^{-3} \end{aligned}$$

$$\frac{\Delta V}{V} = e_{11} + e_{22} + e_{33} \rightarrow$$

$$\Delta V = V(e_{11} + e_{22} + e_{33})$$

$$= \frac{4}{3}\pi(r)^3(e_{11} + e_{22} + e_{33})$$

$$\frac{4}{3}\pi(5)^3(0.033 - 0.0232 - 0.0126) \times 10^{-3} = 3.035 \times 10^{-3}\text{cm}^3$$

5. Figure shows two rectangular cubes with dimensions $a \times b \times c$ that are separated by a rigid plate. A constant pressure of P is applied from upside and the cubes are fixed in another plate with rigid wall. If Poisson's ratio is ν and Young module is E, find the principal stresses as well as the strains for the cubes.

$$\text{Cube1}\begin{cases} \sigma_x^{(1)} = 0 \\ \sigma_y^{(1)} = ? \\ \sigma_z^{(1)} = -P \end{cases}$$

$$\text{Cube2}\begin{cases} \sigma_x^{(2)} = 0 \\ \sigma_y^{(2)} = ? \\ \sigma_z^{(2)} = -P \end{cases}$$

$\sigma_y^{(1)} = \sigma_y^{(2)}$ due to symmetry but the total strain along y direction is zero for both cubes because the walls pretend any strain and fix the cubes.

$$\sigma_y^{(1)} + \sigma_y^{(2)} = 0 \rightarrow \frac{1}{E}\left[\sigma_y^{(1)} - \nu\left(\sigma_X^{(1)} + \sigma_Z^{(1)}\right)\right] + \frac{1}{E}\left[\sigma_y^{(2)} - \nu\left(\sigma_X^{(2)} + \sigma_Z^{(2)}\right)\right]$$

$$\equiv 0 \rightarrow \sigma_y - \nu\sigma_Z^{(1)} + \sigma_y - \nu\sigma_Z^{(2)} = 0$$

$$\rightarrow \sigma_y = \frac{-\nu(P + P)}{2} = -\nu P$$

$$\text{Cube 1} : \sigma_x^{(1)} = 0, \quad \sigma_y^{(1)} = -P\nu, \quad \sigma_z^{(1)} = -P$$

$$\text{Cube 2} : \sigma_x^{(2)} = 0, \quad \sigma_y^{(2)} = -P\nu, \quad \sigma_z^{(2)} = -P$$

$$\text{Srains of Cube 1:} \begin{cases} e_x^{(1)} = \frac{1}{E}\left[\sigma_x^{(1)} - v\left(\sigma_y^{(1)} + \sigma_z^{(1)}\right)\right] = \frac{1}{E}[0 - v(-Pv - P)] = \frac{Pv^2 + Pv}{E} \\ e_y^{(1)} = \frac{1}{E}\left[\sigma_y^{(1)} - v\left(\sigma_x^{(1)} + \sigma_z^{(1)}\right)\right] = 0 \\ e_z^{(1)} = \frac{1}{E}\left[\sigma_z^{(1)} - v\left(\sigma_x^{(1)} + \sigma_y^{(1)}\right)\right] = \frac{1}{E}[-P - v(0 + -Pv)] = \frac{Pv^2 - P}{E} \end{cases}$$

The values of $e_x^{(2)}$, $e_y^{(2)}$, $e_z^{(2)}$ are similar for both rectangular cubes. Now, we calculate the strains which are similar for both rectangular cubes:

$$\frac{\Delta a}{a} = ex \rightarrow \Delta a = ae_x = \frac{a\left(Pv^2 + Pv\right)}{E}$$

$$\frac{\Delta b}{b} = ey \rightarrow \Delta b = be_y = 0$$

$$\frac{\Delta c}{c} = ez \rightarrow \Delta c = ce_z = \frac{c\left(Pv^2 - P\right)}{E}$$

Applications of Continuum Mechanics

6

In this chapter, we will focus on the applications of continuum mechanics in biomedical issues. Continuum mechanics is one of the known methods for the exact description of biological systems due to its ability to calculate stresses and strains. We discuss below the continuum mechanics applications in soft tissues, skin, cartilages, dental implants, cardiovascular system, and biofluids.

Navier–Stokes Equations

The Navier–Stokes equations are very important in fluid mechanics and their application in biomedical engineering. To model blood flow in the cardiovascular system and understand airflow in the respiratory system, the Navier–Stokes equations are widely used along with computational fluid dynamics and fluid–structure interaction. The commercial software, such as COMSOL, Abaqus, and ANSYS-Fluent, is very powerful for analysing fluid mechanics problems, including blood flow in the arteries. They use the Navier–Stokes equations as one of the fundamental codes for solving biofluid problems. There are some examples in the field of flow analysis using continuum mechanics tensors:

1. Blood flow and valve leaflets:

The FSI modelling is normally used to investigate the numerical interaction between blood flow and the valve leaflets. The blood flow is governed by the incompressible Navier-Stokes and continuity equations as follows:

$$\frac{\partial v}{\partial t} + v \cdot \nabla v = -\frac{\nabla p}{\rho} + \nu \nabla^2 v + b, \ \nabla \cdot v = 0 \tag{6.1}$$

K. Hassani and F. Azimifar, *Introduction to Continuum Mechanics*, Synthesis Lectures on Mechanical Engineering, https://doi.org/10.1007/978-3-032-05076-2_6

where v is the flow velocity vector, t is time, ρ is the mass density, p is the pressure, ν is the kinematic viscosity, and b is the body force vector

The leaflet movement can be defined based on Newton's second law as:

$$M\ddot{\theta} = I\ddot{\theta} \tag{6.2}$$

where M = MP + MV, and MP is applied on the leaflets by the pressure field, and MV is due to the viscous forces. The following equation is used by the ANSYS software to iterate the numerical solving:

$$\ddot{\theta}_{k+1} = \ddot{\theta}_k + w\left(\frac{M}{I} - \ddot{\theta}_k\right) \tag{6.3}$$

ω mentions the under-relaxation factor. The equation is iterated until convergence is achieved.

In some special studies in the field of cardiovascular fluid biomechanics, the mass transportation equations between blood and vapor phases can be defined as:

$$\begin{aligned} &\frac{\partial}{\partial t}[(1-\alpha)\rho_l] + \nabla\left[(1-\alpha)\rho_l\vec{V}\right] = -R \\ &\frac{\partial}{\partial t}(\alpha\rho_v) + \nabla\cdot\left(\alpha\rho_v\vec{V}\right) = R \end{aligned} \tag{6.4}$$

α is used for measuring the volume fraction of the vapor phase in the fluid, ρ_l, and ρ_v are densities of the liquid and vapor phase, V is the field velocity that affects on the movement of both phases and their interactions within the domain of the flow, R is the source term for mass transfer between liquid and vapor phases.

2. Blood flow in arteries:

Two methods for analysing the blood flow inside the arteries are Computational fluid dynamics (CFD) and Fluid-structure interaction (FSI). CFD calculates hemodynamic parameters like velocity and pressure using numerical methods and correlates them with the mass and momentum of blood flow in the arteries. Navier-Stokes, Continuity equation, and conservation of mass and energy are the main equations that are used for different CFD numerical simulations. It should be noted that CFD normally uses rigid walls as well as Newtonian blood flow:

$$\begin{aligned} &\rho\left(\frac{\partial \vec{v}_f}{\partial t} + \vec{v}_f\cdot\nabla\vec{v}_f - \vec{f}\right) = \nabla\cdot\vec{\sigma} - \nabla \mathrm{p} \\ &\nabla\cdot\vec{v}_f = 0 \end{aligned} \tag{6.5}$$

v_f is the velocity vector of the fluid, and f is the body force vector per unit volume acting in the direction of fluid flow. The Cauchy stress tensor is σ and p is the acting pressure, while ρ is the fluid mass density.

FSI uses CFD and Finite element method (FEM) and is used for artery wall deformations. It uses a Lagrangian movable mesh for the structural problem and a Lagrangian–Eulerian fixed computational mesh for the fluid domain. Thus, the incompressible Navier-Stokes equation is defined as:

$$\rho\left[\frac{\partial \underset{f}{\vec{v}}}{\partial t}+\left(\underset{f}{\vec{v}}-\underset{grid}{\vec{v}}\right)\cdot\nabla\underset{f}{\vec{v}}-\vec{f}\right]=\nabla^2\mu\underset{f}{\vec{v}}-\nabla \mathrm{p} \tag{6.6}$$

where v_{grid} is the velocity of the moving mesh, μ is the dynamic viscosity ρ is the blood density. The elastic equation for vessel wall deformation is given by:

$$\nabla\cdot\underset{s}{\vec{\sigma}}=\rho_s\left(\frac{d^2\vec{y}}{dt^2}+\eta\frac{d\vec{y}}{dt}+\underset{s}{\vec{f}}\right) \tag{6.7}$$

which σ s represents the Cauchy stress tensor, y is the structural displacement, ρs is the solid wall density, η is the damping coefficient, and fs is the solid domain external body force.

One of the interesting relations that is derived from continuum mechanics relations and is used for the calculation of the arterial wall shear stress is:

$$T_{wall}=\mu\cdot\left[(\nabla\times u)\cdot\vec{n}\right] \tag{6.8}$$

where μ represents the dynamic viscosity, u is the fluid velocity, and $\vec{n}$ Denotes the normal unit vector to the plane tangential to the artery's wall.

3. Deformation of a red blood cell in a capillary:

The red cell membrane, approximately incompressible, has two fat layers which enable the cell to tolerate big deformations or extreme stretch. In this case, the cell shows an elastic response to the applied shear stresses. In this regard, the membrane could be modeled by a hyperelastic material. The following relations are suggested for cell deformation:

$$U=C_{10}(\bar{I}_1-3)+C_{30}(\bar{I}_1-3)^3 \tag{6.9}$$

$$\bar{I}_1=\frac{I_1}{J^{1/3}} \tag{6.10}$$

$$I_1=\lambda_1^2+\lambda_2^2+\lambda_3^2 \quad , \quad J=\lambda_1\lambda_2\lambda_3 \tag{6.11}$$

U is the energy function where I_1 and J are the strain eigenvalues. Relation 10 refers to the strain deviation. $\lambda_1, \lambda_2, \lambda_3$ are the eigenstrains. C_{10}, C_{30} are material constants that are related to the membrane's shear modulus. The viscoelastic effect on the hyperelastic membrane was determined by viscous shear stress (σ_{ij}^{v}) and elastic shear stress (σ_{ij}^{e}). The following tensor represents the membrane shear vectors [15, 16]:

$$\sigma_{ii}^{s} = \sigma_{ii}^{e} + \sigma_{ii}^{v} \tag{6.12}$$

$$\sigma_{ii}^{e} = \left(\frac{\xi}{2}\right)\left(\lambda_i^2 - 1\right), \quad \sigma_{ii}^{v} = 2\eta\left(\frac{\dot{\lambda}_i}{\lambda_i}\right) \tag{6.13}$$

ξ Is the membrane shear module and η It is the viscosity of the membrane surface. The blood flow was assumed as a suspension of plasma and platelet particles without white cells. The white cells were neglected due to their very low concentration. The following experimental relation was proposed for the effective viscosity of the suspension:

$$\mu_{eff} = \mu_{bf}(53.742\varphi - 0.188) \tag{6.14}$$

where μ_{bf} is the basic fluid viscosity and φ Is the volume concentration. The blood density was determined by the following equation:

$$\rho_{eff} = (1 - \varphi)\rho_f + \varphi\rho_p \tag{6.15}$$

where ρ_f Is the density of the basic fluid and ρ_p It is particle density. Three main domains are required for numerical modeling of the red cell motion and its deformation in the capillary, including calculation of blood suspension, the cell layer, and the cell cytoplasm. The capillary was assumed as a pipe with a length of 60 μm and a diameter of 12 μm. The initial location of the cell was 10 μm from the inlet, and the cell was positioned horizontally, vertically, and angular (45 degrees. The following relation is obtained for the two-dimensional profile of the red cell:

$$Z = \pm 0.5R_0\left[1 - \frac{X^2 + Y^2}{R_0^2}\right]^{1/2}\left[C_0 + C_1\frac{X^2 + Y^2}{R_0^2} + C_2\left(\frac{X^2 + Y^2}{R_0^2}\right)^2\right] \tag{6.16}$$

$$R_0 = 3.91\,\mu\text{m}, \quad C_0 = 0.207161, \quad C_1 = 2.002558 \text{ and } C_2 = -1.122762$$

R_0 refers to the average radius of the cell. The values of X and Y could be between 7 to 8.5 μm based on the cell dimensions. The cell layer was assumed as a thin wall material with a thickness of 0.27 μm. The blood suspension was considered transient, incompressible, and viscous. The flow behavior inside the capillary is laminar. To solve the numerical problem, including deformation and motion of the cell, the main equations, including the structural equations of the layer, governing equations of the fluid, the motion The governing equations of fluid domain, including continuity and momentum, were solved in two regions: the area inside the capillary which has surrounded the cell and the one inside the cell which has been surrounded by the cytoplasm:

$$\frac{\partial u_j}{\partial x_i} = 0$$
$$\rho_N \frac{\partial u_i}{\partial t} + \rho_N \frac{\partial u_j u_i}{\partial x_i} = -\frac{\partial P}{\partial x_i} + \mu_N \frac{\partial}{\partial x_i}\left(\frac{\partial u_i}{\partial x_j} + \frac{\partial u_j}{\partial x_i}\right) + \rho_N B_i, \quad i,j = 1,2,3 \tag{6.17}$$

$$\mu_N = \begin{cases} \mu_{eff} & blood\ suspension \\ \mu_{citoplasm} & citoplasm \end{cases}, \quad \rho_N = \begin{cases} \rho_{eff} & blood\ suspension \\ \rho_{citoplasm} & citoplasm \end{cases} \tag{6.18}$$

B_i Represents the external force, which is considered only for the flow inside the cell, and its value is zero for the fluid flow inside the capillary. The following boundary conditions were applied to the fluid structure interaction areas in the cell layer:

$$\begin{aligned} &kinematics\ condition \rightarrow d_f = d_s \\ &dynamic\ condition \rightarrow n.\sigma_f = n.\sigma_s \end{aligned} \tag{6.19}$$

where d_f and d_s are the displacements of the fluid and solid. σ_f, σ_s Represent the fluid and solid stresses, and n is the normal vector to the interface area. The move of the central mass is expressed based on Newton's inertial Equation:

$$F^i = F_a^i + F_e^i + F_g^i = m\ddot{r}_{c.m.}^i \tag{6.20}$$

F_a^i Refers to the Hydrodynamic force, F_g^i gravitational force, and F_e^i other forces. The central mass position of the cell, as a function of time, is determined by the following equation.

$$r_{c.m.}^i(t^{n+1}) = \frac{1}{2}\frac{F^i}{m}\Delta t^2 + u_{c.m.}^i(t^n)\Delta t + r_{c.m.}^i(t^n) \tag{6.21}$$

$u_{c.m}^i$ is the speed of the central mass.

Equations for the six-degree cell and coupled fluid–structure interaction equations are used.

4. Dental Implants:

Permanent implants are an essential part of dentistry surgery, and they act as an effective substitute for lost teeth. Statistics show that implants are installed widely all around the world. Although the implant technology has advanced during the last decade, the problem of implant loss remains a big issue. Due to different biomechanical stresses, the useful life of the implants is around 10 to 15 years. The three-dimensional design of an implant is a crucial issue, including all the composing elements. A dental element is always under different loads, which are applied in various directions. Therefore, an optimum design is needed to manage all biomechanical stresses that are applied to the surrounding tissues.

The stress distribution at the surrounding bone is affected by many factors, such as the implant, which increases the induced bone stress under axial or inclined loadings. Regarding biomechanics, stress shielding is one of the most important issues with existing implants when too many stresses are induced on the implants and it reduces bone strength. one of the primary concerns with commercially available standardized implants is stress-shielding, a phenomenon where the implant absorbs too much load, thus reducing the mechanical stimulation that the bone receives. To study the rate of the induced stress on the implant, Finite element methods can calculate stress distributions and help to find an optimum method of reducing the existing loss of bone-implant interface. Von Mises stress distribution in cancellous bone for integrated and two-piece implants is an important factor. The maximum stress for the interface area between the bone and the two-piece implant is another point. In the next step, von Mises and principal stress tensors of the integration point for both implants in the mandible should be calculated. The von Mises stress values are obtained using available engineering software like ANSYS or Abaqus. But average normal stress, Octahedral stress, and deviation stress tensor for a two-piece implant can be calculated as per the following relations:

The average normal stress:

$$\sigma_m = \frac{1}{3}(\sigma_{11} + \sigma_{22} + \sigma_{33}) \tag{6.22}$$

The normal and shear octahedral stresses:

$$\hat{\sigma} = \frac{1}{3}(\sigma_1 + \sigma_2 + \sigma_3) \tag{6.23}$$

$$\hat{\tau} = \frac{1}{3}\left[(\sigma_1 + \sigma_2)^2 + (\sigma_2 + \sigma_3)^2 + (\sigma_3 + \sigma_1)^2\right]^{1/2} \tag{6.24}$$

The deviation stress tensor:

$$S_{ij} = \sigma_{ij} - \sigma_m \delta_{ij} \tag{6.25}$$

where σ_{ij} is the von Mises stress tensor.

5. Soft tissues:

The soft issues, such as skin, cartilage, Tendons, and muscles, have unique properties which distinguish them from solid materials. They have non-linear, viscoelastic, and anisotropic behavior. Therefore, some new continuum mechanics models, including hyper-elastic (skin) and viscoelastic (cartilage), can be helpful to simulate and model these soft tissues. The cartilage is a porous, viscoelastic tissue that absorbs the vibrations and shocks and distributes the applied stress uniformly. The mechanical behavior of the cartilages depends on the rate of loading and the amount of water inside the cartilage. Thus, viscoelastic and two-phase models are used for modelling.

The skin has a multilayer structure with complicated mechanical properties. It has collagen and elastin, which give an elastic property to the skin. The skin can be considered as a hyperelastic and anisotropic material, which is modelled using Ogden, Mooney-Rivlin, or Neo-Hookean models. The muscles are always under dynamic loads and deform. The contractile and elastic characteristics of the muscles should be considered for modelling their function. They are normally simulated as hyperelastic material and continuum mechanics laws are used for studying the effects of external forces on the deformation and internal stresses inside the muscles.

Supplementy Problems

1. Find the result of each indicial notation sentence:
   (a) $\delta_{ij}.V_j$
   (b) $\delta_{ij}.C_{ij}$
   (c) $\delta_{ij}.y_i.y_i$
   (d) $\delta_{ij}.\delta_{ij}$
   (e) $\delta_{ij}.\delta_{ik}.\delta_{jk}$
   (f) $\left(\delta_{ij} + C_{ij}\right)(\delta_{ik} + C_{ik}) - \delta_{jk} - \delta_{mn}.C_{mj}.C_{nk}$
2. Show that determinant of a 3 × 3 matrix, a_{ij}, can be presented as below, e is permutation symbol.

$$\left|a_{ij}\right| = e_{ijk} \cdot a_{1i} \cdot a_{2j} \cdot a_{3k}$$

3. Prove the following sentences if e is the permutation symbol.
   (1) $e_{ijk}.e_{imn} = \delta_{jm}.\delta_{kn} - \delta_{jn}.\delta_{km}$
   (2) $e_{ijk}.e_{kji} = -6$
   (3) $e_{kks} = 0$
   (4) $e_{ijk}.A_j.A_k = 0$
4. Prove the following using the index notation method. A is a scalar function and F is a position vector. Moreover: $\left(\nabla.\nabla = \nabla^2\right)$

$$\nabla.\left(A\vec{F}\right) = (\nabla A) \times \vec{F} + A\left(\vec{\nabla} \times \vec{F}\right)$$

$$\nabla \times (\nabla A) = 0$$

$$\nabla.\left(\nabla \times \vec{F}\right) = 0$$

K. Hassani and F. Azimifar, *Introduction to Continuum Mechanics*, Synthesis Lectures on Mechanical Engineering, https://doi.org/10.1007/978-3-032-05076-2

$$\vec{\nabla} \times \left(\vec{\nabla} \times \vec{F}\right) = \vec{\nabla}\left(\vec{\nabla}.\vec{F}\right) - \nabla^2\vec{F}$$

5. If r is a position vector, $r = x = x_i.e_i$, in which $r^2 = x_i.x_i$ and A is a scalar, then prove the followings (n is integer):

$$\nabla^2\left(r^n\right) = n(n+1).r^{n-2}$$

$$\nabla(r.A) = A$$

$$\nabla.(r \times A) = 0$$

$$\nabla \times (r \times A) = -2A$$

$$\nabla.(rA) = \frac{1}{r}(r.A)$$

$$\nabla \times (rA) = \frac{1}{r}(r \times A)$$

6. Calulate the eigne values and eigen vectors of the following matrix:

$$T = \begin{bmatrix} -2 & 2 & 10 \\ 2 & -11 & 8 \\ 10 & 8 & -5 \end{bmatrix}$$

7. Calculate the principal stress, principal directions, maximum shear stress, deviation stress, and spherical stress tensor for below stress tensor:

$$\begin{bmatrix} \frac{1}{4} & -\sqrt{6} & -\sqrt{\frac{3}{4}} \\ -\sqrt{6} & 2 & -\sqrt{2} \\ -\sqrt{\frac{3}{4}} & -\sqrt{2} & \frac{3}{4} \end{bmatrix}$$

8. Solve the following items for the stress tensor:

$$\begin{bmatrix} 1 & -3 & -\sqrt{2} \\ -3 & 1 & -\sqrt{2} \\ \sqrt{2} & -\sqrt{2} & 4 \end{bmatrix}$$

A. The stress vector of the plane with normal vector $2i_1 - 2i_2 + i_3$
B. The angle between the vector stress and normal vector of A
C. The values of normal stress and shear stress of the stress vector

9. For the stress tensor: $\begin{bmatrix} 3 & 5 & 8 \\ 5 & 1 & 0 \\ 8 & 0 & 2 \end{bmatrix}$, find the Octahedral stress vector, components of normal and shear values of the Octahedral stress vector.
10. For the state of stress below, obtain the requested items.

$$\sigma_{11} = -2x_1^2, \sigma_{12} = -7 + 4x_1x_2 + x_3, \sigma_{13} = 1 + x_1 - 3x_2$$
$$\sigma_{22} = 3x_1^2 - 2x_2^2 + 5x_3, \sigma_{23} = 0, \sigma_{33} = -5 + x_1 + 3x_2 + 3x_3$$

The stress vector applied on point $$\left({x_{1} ,x_{2} ,x_{3} } \right)$$ which belongs to the plane: $$x_{1} + x_{2} + x_{3} = constant$$

The normal and shear components of the stress vector applied at point (1,1,3)

The principal stress and principal directions at point (1,2,1)

11. For the stress tensor $\begin{bmatrix} 5x_2x_3 & 3x_2^2 & 0 \\ 3x_2^2 & 0 & -x_1 \\ 0 & -x_1 & 0 \end{bmatrix}$, find the volume forces in equilibrium condition of the environment.
12. For the stress tensor applied at point P, $\begin{bmatrix} 1 & 2 & 1 \\ 2 & \sigma_0 & 0 \\ 1 & 0 & -3 \end{bmatrix}$, find the normal vector of passing plane from P that the point is in zero stress condition. Find σ_0 also in this condition.
13. The motion of a continuum body has given by below equations, find:

$$\begin{cases} x_1 = \left(1 + e^t\right)X_1 \\ x_2 = \left(1 + e^{-2t}\right)X_2 \\ \mathrm{x}_3 = \mathrm{X}_3 \end{cases}$$

   A. The gradient tensor F
   B. The velocity components in Lagrangian and Eulerian coordinates.
   C. The acceleration components in Eulerian coordinates.
14. For below given deformation functions, find:

$$\begin{cases} x_1 = X_1 + Ax_2 \\ x_2 = X_2 + Bx_1 \\ x_3 = X_3 \end{cases}$$

   A. The displacement function, u, components in Lagrangian and Eulerian coordinates.
   B. The Green–Lagrange and Almansi strain tensors
15. The motion of a continuum body is defined by following functions, find:

$$\begin{cases} x_1 = \frac{1}{2}(X_1 + X_2)e^t + \frac{1}{2}(X_1 - X_2)e^{-t} \\ x_2 = \frac{1}{2}(X_1 + X_2)e^t - \frac{1}{2}(X_1 - X_2)e^{-t} \\ x_3 = X_3 \end{cases}$$

A. The velocity components in Lagrangian and Eulerian coordinates
B. The rate of deformation tensor, D_{ij}, and vorticity tensor W_{ij}

16. The rectangle ABCD is deformed under following motion equations, obtain displacement functions(u) and Green–Lagrange strain tensor.

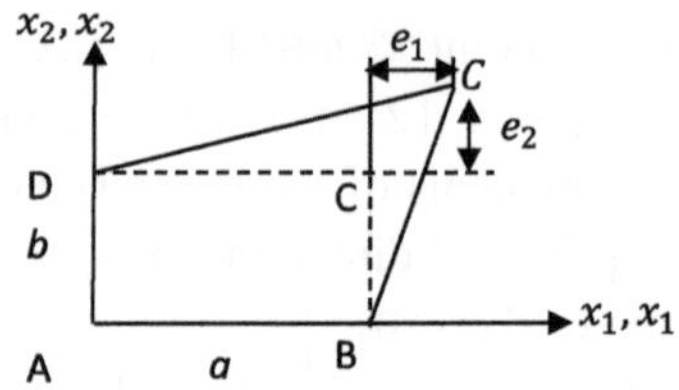

$$\begin{cases} x_1 = a_1X_1 + a_2X_2 + a_3X_1X_2 \\ x_2 = b_1X_1 + b_2X_2 + b_3X_1X_2 \\ x_3 = X_3 \end{cases}$$

17. For the motion field, find:

$$\begin{cases} x_1 = 3.5 + X_1 + \frac{X_2}{2} \\ x_2 = 4 + X_2 \\ x_3 = X_3 \end{cases}$$

A. The deformation gradient F
B. Green–Lagrange strain tensor
C. The deformed volume.
D. The deformed surface vector if the initial surface vector is i_2.

18. Figure shows a rectangular bock, $3b \times 4b \times h$, with a triangular hole inside it. If the block is under deformation of following field, then find:

$$\begin{cases} x_1 = X_1 + X_2 \\ x_2 = X_2 \\ x_3 = X_3 \end{cases}$$

A. The new angle value $A\hat{B}C$ for deformed condition.
B. The deformed surface magnitude for ABC triangle.

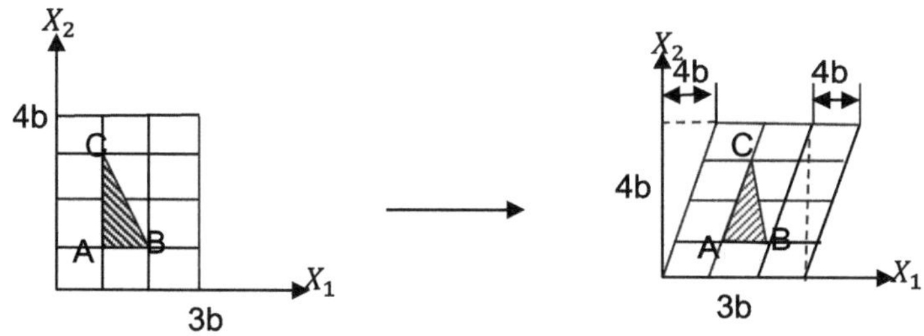

19. For the motion field of a continuum body with following equations, find the density (ϱ) according to x and t.

$$\begin{cases} x_1 = \frac{X_1}{1+tX_1} \\ x_2 = X_2 \\ x_3 = X_3 \end{cases}$$

20. For the stress tensor below, find the volume forces components to let the continuum body be in equilibrium position.

$$\sigma = \begin{bmatrix} x_1x_2 & x_1^2 & -x_2 \\ x_1^2 & 0 & 0 \\ -x_2 & 0 & x_1^2 + x_2^2 \end{bmatrix}$$

21. For an isotropic under two-dimensional stress that the displacement functions are as below, find the stress tensor at (0.05, 0.02 m). $E = 210Gpa, v = 0.3$

$$\begin{cases} u_1 = 3x_1^2 - x_1^3x_2 + 2x_2^3 \\ u_2 = x_1^3 + 2x_1x_2 \end{cases}$$

22. For an isotropic body under stress, the stress tensor at a point is given by below, find:

$$\sigma = \begin{bmatrix} 6 & 2 & 0 \\ 2 & -3 & 0 \\ 0 & 0 & 0 \end{bmatrix}$$

A. The engineering strain tensor for $G = 80^{Gpa}$, $E = 207^{Gpa}$
B. The deformed volume of a 5 cm cube.

23. For displacement functions given below and assuming small strains, find:
A. The Eulerain strain tensor and rotation tensor.
B. Prinicpal strain

$$\begin{cases} u_1 = x_1^2 - 5x_2^2 \\ u_2 = 2x_1x_2 \\ u_3 = 0 \end{cases}$$

24. Find the strain tensor for the stress tensor of an isotropic material with lame coefficients $\mu = 20000kpa$, $\lambda = 20000kpa$

$$\sigma = \begin{bmatrix} -26 & 6 & 0 \\ 6 & 9 & 0 \\ 0 & 0 & 29 \end{bmatrix}$$

25. A point of an isotropic body is under the below strain tensor. For $E = 10Mpa$, $\nu = 0.25$, find:

$$e_{ij} = \begin{bmatrix} -2 & 3 & 0 \\ 3 & -10 & 0 \\ 0 & 0 & 0 \end{bmatrix} \times 10^{-3}$$

The volume strain.
The deviator strain tensor.
The related stress tensor.

26. For the displacement field given below, find the related stress and strain tensors. Investigate what the condition for equilibrium of the continuum is if there are no volume forces.

$$\begin{cases} u_1 = u_2 = 0 \\ u_3 = sinb(x_3 - ct) + asinb(x_3 + ct) \end{cases}$$

Multiple Choice Problems

1. If the eigen vector and eigen value of matrix A are x and λ, find the eigen vector and eigen value of A^5.
   (1) 5λ, x
   (2) $\lambda, 5x$
   (3) λ, x^5
   (4) λ^5, x
2. Bij is an antisymmetric matrix. find $\varepsilon_{pqi} b_i$ if the vector $b_i = \frac{1}{2} \varepsilon_{ijk} B_{jk}$.
   (1) $\frac{1}{2} B_{pq}$
   (2) B_{pq}
   (3) $2B_{pq}$
   (4) $\delta_{ij} B_{pq}$
3. What is the normal stress component vector for a plane with normal vector n, and if the coordinate axes are in the direction of principal directions?$n = \frac{1}{\sqrt{3}}(e_1 + e_2 - e_3)$

$$\begin{bmatrix} \sigma_1 & 0 & 0 \\ 0 & \sigma_2 & 0 \\ 0 & 0 & \sigma_3 \end{bmatrix}, \ I_1 = \sigma_1 + \sigma_2 + \sigma_3$$

   (1) $\frac{I_1}{3\sqrt{3}}(e_1 + e_2 - e_3)$
   (2) $\frac{I_1}{3}(e_1 + e_2 - e_3)$
   (3) $\frac{I_1}{\sqrt{3}}(e_1 + e_2 - e_3)$
   (4) $\sqrt{3} I_1 (e_1 + e_2 - e_3)$
4. If V is a vector field, what is the result of $\nabla \times V$

$$V = \frac{-x_2 i_1 + x_1 i_2}{\left(x_1^2 + x_2^2\right)}$$

K. Hassani and F. Azimifar, *Introduction to Continuum Mechanics*, Synthesis Lectures on Mechanical Engineering, https://doi.org/10.1007/978-3-032-05076-2

(1) $\frac{-i_1+i_2}{(x_1^2+x_2^2)^2}$
(2) $\frac{i_1-i_2}{(x_1^2+x_2^2)^2}$
(3) $\frac{-i_1+i_2}{(x_1^2+x_2^2)}$
(4) 0

5. The deformation field of a continuum environment is as below. Find the deformed line vector direction if the initial line vector direction was $i_1 + i_2 + i_3$

$$x_1 = \sqrt{2}X_1 + \frac{3\sqrt{2}}{4}X_2, x_2 = -X_1 + \frac{3}{4}X_2 + \frac{\sqrt{2}}{4}X_3, x_3 = X_1 - \frac{3}{4}X_2 + \frac{\sqrt{2}}{4}X_3$$

(1) $\frac{1}{4}\begin{bmatrix} 7\sqrt{2} \\ \sqrt{2}-1 \\ \sqrt{2}+1 \end{bmatrix}$
(2) $\frac{1}{4}\begin{bmatrix} -7\sqrt{2} \\ \sqrt{2}-1 \\ \sqrt{2}+1 \end{bmatrix}$
(3) $\frac{1}{4}\begin{bmatrix} 7\sqrt{2} \\ \sqrt{2}+1 \\ \sqrt{2}-1 \end{bmatrix}$
(4) $\frac{1}{4}\begin{bmatrix} -7\sqrt{2} \\ \sqrt{2}+1 \\ \sqrt{2}-1 \end{bmatrix}$

6. The Eulerian velocity field in a continuum environment is given below.Find the rate of extension, $D_{ij}.n_i.n_j$, for the direction $\vec{e}_1 + \vec{e}_2$. Note that α, β, γ are constant.

$$v_1 = \alpha x_2 - \beta x_3, v_2 = \beta x_3 - \gamma x_1, v_3 = \gamma x_1 - \alpha x_2$$

(1) $\gamma - \alpha$
(2) $\alpha - \Upsilon$
(3) $\frac{\alpha-\gamma}{2}$
(4) $\frac{\gamma-\alpha}{2}$

7. Assume the velocity components of a continuum body are as below. Find the parametric value of the density ρ according to the continuity equation $\frac{D\rho}{Dt} + \rho\nabla.v = 0$

$$v_2 = v_3 = 0, v_1 = -\frac{X_1^2}{(1-tX_1)^2} = -x_1^2(x_1, X_1, t > 0)$$

(1) $\frac{\rho_\circ}{\rho_\circ(1-tx_1)}$
(2) $\frac{\rho_\circ}{(1-tx_1)^2}$
(3) $\rho_\circ(1-tx_1)$
(4) $\rho_\circ(1-tx_1)^2$

8. In following the stress–strain relation, which sentence is true when i = j?

$$\sigma_j^i = \frac{E}{1+v}\left(E_j^i + \frac{v}{1-2v}E_m^m\sigma_j^i\right)$$

(1) $\frac{E}{(1+2v)}E_m^m$
(2) $\frac{E}{1-2v}E_m^m$
(3) $\frac{(1-v)E}{(1+v)(1-2v)}E_m^m$
(4) $\frac{(1+v)E}{(1-v)(1+2v)}E_m^m$

9. For below stress tensors, what is the ratio of maximum shear stress of tensor b to a?

(a) $\sigma_{ij} = \begin{bmatrix} \tau & \tau & 0 \\ \tau & \tau & 0 \\ 0 & 0 & 0 \end{bmatrix}$

(b) $\sigma_{ij} = \begin{bmatrix} \tau & 0 & 0 \\ 0 & -\tau & 0 \\ 0 & 0 & 2\tau \end{bmatrix}$

(1) $\frac{2}{3}$
(2) $\frac{3}{4}$
(3) $\frac{4}{3}$
(4) $\frac{3}{2}$

10. If the Eulerian velocity field of a continuum environment is as below, what is the acceleration of a particle at (x_1, x_2, x_3)

$$V_1 = Ax_2, V_2 = Ax_1, V_3 = B$$

(1) $-A^2(X_1, X_2, 0)$
(2) $A^2(X_1, X_2, 0)$
(3) $-A(X_1, X_2, 0)$
(4) $A(X_1, X_2, 0)$

11. The deformation gradient of a continuum body is given as $\vec{F} = \begin{bmatrix} a & d & 0 \\ 0 & b & 0 \\ 0 & 0 & c \end{bmatrix}$. Which relation should be between the parameters to keep the initial surface vector e_1 unchanged after deformation?

(1) $abc = 1$
(2) $a^2 + b^2 = c^2d^2$
(3) $b^2 + d^2 = a^2b^2$
(4) $b^2c^2 + c^2d^2 = 1$

12. The Cauchy stress tensor for a point of a continuum body is given below. Which plane has a zero normal stress value?

$$\vec{T} = \begin{bmatrix} 100 & 100 & 0 \\ 100 & -300 & 0 \\ 0 & 0 & 200 \end{bmatrix} \text{MPa}$$

(1) $x_1 + x_2 = 2$
(2) $x_1 + x_2 = 4$
(3) $x_1 + x_3 = 4$
(4) $x_1 + x_3 = 2$

13. If δ_{ij} is the Kronecker delta and ε_{ijk} is the permutation symbol, find $\varepsilon_{ijk}\delta_{km}\varepsilon_{mji}$
(1) 0
(2) 1
(3) 6
(4) −6

14. In following the Eulerian velocity field, find the ratio of extension, $D_{ij}.n_i.n_j$, for a line which makes equivalent angles with the Cartesian coordinate axes.
$v_1 = x_1 - 2x_2, v_2 = x_2 - 2x_1, v_3 = 0$
(1) $-\frac{1}{3}$
(2) $-\frac{4}{3}$
(3) $-\frac{5}{3}$
(4) $-\frac{2}{3}$

15. The motion of a body is given as below. if the Eulerian temperature field is defined as $\theta = x_1 + x_2$ what is the rate of temperature variation $\frac{D\theta}{Dt}$?

$$x_1 = X_1 + ktX_2, x_2 = X_2, x_3 = X_3$$

(1) kX_1
(2) kX_2
(3) kX_2
(4) $X_1 + (kt + 1)X_2$

16. What are the values of K to let the deformation be possible for the following motion field: $x_1 = X_1 - kX_3, x_2 = X_2 + KX_1, x_3 = X_3 - kX_2$
(1) $(-\infty, -1)$
(2) $(-1, +\infty)$
(3) $(-\infty, -1]$
(4) $[-1, +\infty)$

17. The velocity field of a Newtonian fluid is given below in Eulerian coordinates. What is the relation between constants k and c for an incompressible fluid?

$$v = \begin{Bmatrix} kx_1 \\ -3kx_2 \\ cx_3 \end{Bmatrix}$$

(1) $c = k$
(2) $c = 2k$
(3) $c = 3k$
(4) $c^2 + 10k^2 = 1$

18. If $S = A_{ij}X_iX_j$ and A_{ij} is a constant value, what is the value of $\frac{\partial S}{\partial X_i}$
(1) $2A_{ij}X_i$
(2) $2A_{ij}X_j$
(3) $2A_{ij}X_i$
(4) $(A_{ij} + A_{ji})X_j$

19. For a special static continuum environment, the stress field is considered by the relation $\sigma_{ij} = -P_\circ\delta_{ij}$ which $P_\circ$ is a constant. What is the value of stress power or $\sigma_{ij}D_{ij}$ if the continuity equation is defined as $\dot{\rho} + \rho v_{i,i} = 0$
(1) $\frac{P_\circ \dot{\rho}}{\dot{\rho}}$
(2) $\frac{P_\circ \rho}{2\dot{\rho}}$
(3) $\frac{P_\circ \dot{\rho}}{\rho}$
(4) $\frac{P_\circ \dot{\rho}}{2\rho}$

20. In a continuum environment, the motion field is given below. What is the component of Eulerian velocity in the x direction of Cartesian coordinates?

$$x_1 = X_1(1 + t)$$

$$x_2 = X_2(1 + t)^2$$

$$x_3 = X_3(1 + t^2)$$

(1) $v_1 = \frac{x_1}{1+t}$
(2) $v_1 = \frac{2x_1}{1+t}$
(3) $v_1 = x_1(1 + t)$
(4) $v_1 = 2x_1(1 + t)$

21. The square OABC, whose side length is unit, is deformed by the displacement field as shown below. What is the deformed value of the angle in the C corner? K is small.

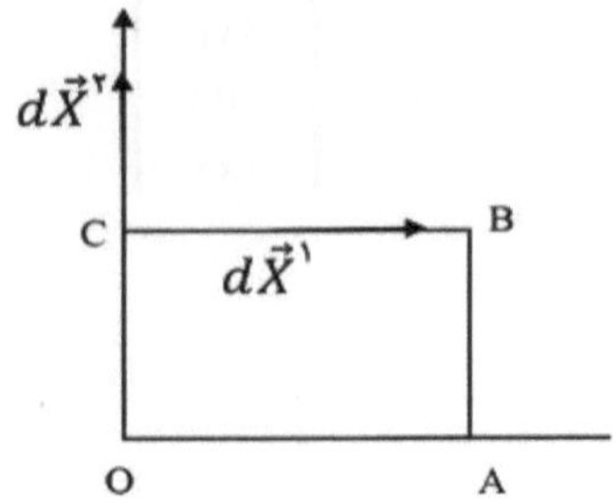

$$u_1 = kX_2^2,\ u_2 = u_3 = 0$$

$$d\vec{X}^1 = dX_1\vec{e}_1,\ d\vec{X}^2 = dX_2\vec{e}_2$$

(1) 0
(2) k
(3) 2k
(4) $\frac{2k}{\sqrt{1+4k^2}}$

22. Hook law is given below for an isotropic elastic material. Find the strain relation.
(1) $\varepsilon_{ij} = \frac{1}{2\mu}\left(\sigma_{ij} - \frac{\lambda}{3\lambda+2\mu}\delta_{ij}\sigma_{kk}\right)$
(2) $\varepsilon_{ij} = \frac{1}{2\mu}\left(\sigma_{ij} + \frac{\lambda}{3\lambda+2\mu}\delta_{ij}\sigma_{kk}\right)$
(3) $\varepsilon_{ij} = \frac{1}{2\mu}\left(\sigma_{ij} - \frac{\lambda}{3\lambda+2\mu}\delta_{ij}\sigma_{ij}\right)$
(4) $\varepsilon_{ij} = \frac{1}{2\mu}\left(1 - \frac{\lambda}{3\lambda+2\mu}\right)\sigma_{ij}$

23. In following the deformation field, what is the relation between the constants α, β for j = 1?

$$x_1 = X_1 + \alpha X_2 + \alpha\beta X_3,\ x_2 = \alpha\beta X_1 + X_2 + \beta^2 X_3,\ x_3 = X_1 + X_2 + X_3,$$

(1) $\beta = \frac{\alpha^2+\alpha}{\alpha^2+\alpha-1}$
(2) $\beta = \frac{\alpha^2+\alpha}{\alpha^2+\alpha+1}$
(3) $\beta = \frac{\alpha^2-\alpha}{\alpha^2-\alpha+1}$
(4) $\beta = \frac{\alpha^2-\alpha}{\alpha^2-\alpha-1}$

24. The Green–Lagrange strain tensor is defined for a point of a continuum environment as below. Find the stretch in the line with direction vector $i_1 – i_2 + i_3$

$$E = \begin{bmatrix} \frac{5}{2} & 0 & 0 \\ 0 & 4 & 0 \\ 0 & 0 & 0 \end{bmatrix}$$

(1) $\frac{16}{3}$
(2) $\sqrt{\frac{2}{3}}$
(3) $\frac{4}{3}$
(4) $\frac{4\sqrt{3}}{3}$

25. The motion of a continuum environment is given in Eulerian coordinates as below. What is the value of Jacobian (j)?

$$x_1 = X_1 e^t + X_3\left(e^t - 1\right)$$

$$x_2 = X_3\left(e^t - e^{-t}\right) + X_2$$

$$x_3 = X_3$$

(1) 1
(2) e^t
(3) e^{-t}
(4) $e^t - 1$

26. What is the total Eulerian acceleration value of a particle with the following motion field in Lagrangian coordinates? k is constant

$$x_1 = (1 + kt)X_1 x_2 = (1 + kt)X_2 x_3 = (1 + kt)X_3$$

(1) 0
(2) $\frac{kx_i}{1+kt}$
(3) $\frac{k^2 x_i}{(1+kt)^2}$
(4) $-\frac{k^2 x_i}{(1+kt)^2}$

27. For the below given motion and temperature fields, find $\frac{D\theta}{Dt}$

$$x(x, t) = (X_1 + \Lambda t X_2)\hat{e}_1 + (X_2 - \Lambda t X_1)\hat{e}_2 + X_3\hat{e}_3$$

$$\theta(x, t) = x_1 + t x_2$$

(1) $X_1 + tX_1$
(2) $\Lambda X_2 - \Lambda X_1$
(3) $\Lambda t X_2 - \Lambda X_1$
(4) $-2\Lambda t X_1 + (1+\Lambda) X_2$

28. For the given motion field, which item is true?

$$x_1 = X_1 e^t$$

$$x_2 = X_1\left(e^t - 1\right) + X_2$$

$$x_3 = X_3$$

(1) The motion is not possible
(2) The motion is possible, and the deformation is uniform
(3) The motion is possible and the deformation occurs in a constant volume condition
(4) The motion is possible, and the deformation is not uniform

29. If there is no volume force and the stress tensor is given as below, which item is true? The body is in an equilibrium condition

$$\left[\sigma_{ij}\right] = \begin{bmatrix} (A-1)x_1 + x_2 & B(x_1 + x_2) & Cx_3 \\ B(x_1 + x_2) & 0 & -Cx_3 \\ Cx_3 & -Cx_3 & Ax_1x_2 + (B+1)x_3 \end{bmatrix}$$

(1) $A = 1, B = 3, C = 1$
(2) $A = -1, B = 3, C = -1$
(3) $A = 3, B = -1, C = -1$
(4) $A = 3, B = 1, C = 1$

30. For a small deformation, α is a very small positive scalar, with the following displacement functions. Find the volume strain $\frac{\Delta V}{V}$

$$\begin{cases} u_1 = \alpha(x_1 + x_2) \\ u_2 = 2\alpha x_2 \\ u_3 = 3\alpha x_3 \end{cases}$$

(1) $\frac{1}{3}\alpha$

(2) 3α
3) 6α
(4) 12α

31. Which item is not equivalent to C_{1232} in the elastic coefficient tensor?
(1) C_{2321}
(2) C_{3221}
(3) C_{2312}
(4) C_{2231}

32. The stress tensor of a point is given by $\begin{bmatrix} 1 & -5 & 1 \\ -5 & 2 & -1 \\ 1 & -1 & 5 \end{bmatrix}$. What is the stress vector of the point in a plane with a normal vector of $\frac{\hat{e}_1}{9} - \frac{8\hat{e}_2}{9} + \frac{4\hat{e}_3}{9}$.
(1) $-9\hat{e}_1 - \frac{29}{5}\hat{e}_2 + 5\hat{e}_3$
(2) $-5\hat{e}_1 - \frac{25}{9}\hat{e}_2 - \frac{29}{9}\hat{e}_3$
(3) $5\hat{e}_1 - \frac{29}{9}\hat{e}_2 + \frac{29}{9}\hat{e}_3$
(4) $9\hat{e}_1 - \frac{29}{5}\hat{e}_2 + \frac{29}{9}\hat{e}_3$

33. For the deformation field given below, find the extension (λ) of the line with the direction vector $\vec{e}_2$.

$$x_1 = X_1 + kX_2$$

$$x_2 = X_2$$

$$x_3 = X_3$$

(1) $\sqrt{1+k^2}$
(2) $2\sqrt{1+k^2}$
(3) k
(4) $1+k^2$

34. If ε_{ijk} is the permutation symbol, find $\varepsilon_{mnp}\varepsilon_{nrm}$.
(1) $-2\delta_{pr}$
(2) 0
(3) $2\delta_{pr}$
(4) 6

35. The principal values of a deviator stress tensor are the roots of the equation below. Find the Octahedral shear stress.

$$S^3 - 294S = 0$$

(1) 14
(2) $14\sqrt{2}$
(3) 21
(4) $21\sqrt{2}$

36. In the equilibrium condition of a continuum environment, the volume forces are equivalent to:
(1) Divergence of stress
(2) Curl of stress
(3) minus sign Curl of stress with
(4) Divergence of stress with minus sign

37. The Cauchy stress tensor of a point $(4, -1, 2\sqrt{2})$ which is on a sphere, is given below. The center of the sphere is confined to the center of the Cartesian coordinates. Find the component of the stress vector of the point in x direction?

$$\sigma = \begin{bmatrix} 1 & -1 & 0 \\ -1 & 2 & 3 \\ 0 & 3 & -1 \end{bmatrix}$$

(1) 5
(2) -1
(3) 1
(4) 5

38. What is the equivalent value of the octahedral normal stress, σ_n^{oct}, based on the first stress invariant I_1.
(1) $\sigma_n^{oct} = \frac{I_1}{\sqrt{3}}$
(2) $\sigma_n^{oct} = \frac{\sqrt{I_1}}{3}$
(3) $\sigma_n^{oct} = I_1$
(4) $\sigma_n^{oct} = \frac{1}{3}I_1$

39. What is the value of $\delta_{ij}\delta_{jk}\delta_{kl}\delta_{Ir}\delta_{ri}$. δ_{ij} is the Kronecker delta.
(1) 1
(2) 3
(3) δ_{ir}
(4) δ_{Ir}

40. Which of the following is true regarding index notation rules?
   (1) $\frac{\partial y_i}{\partial x_j}c_k + \frac{\partial y_j}{\partial x_k}c_i c_i = T_{ii}\frac{\partial s_{jk}}{\partial x_j}$
   (2) $T_n\frac{\partial c_m}{\partial x_m} + z_k y_k e_n = \frac{\partial f_{in}}{\partial x_i}$
   (3) $\frac{\partial a_i}{\partial x_k}\frac{\partial b_j}{\partial x_k} + \frac{\partial y_j}{\partial x_j}c_k = \frac{\partial f_i}{\partial x_k}\frac{\partial g_k}{\partial x_j}$
   (4) $\frac{\partial x_i}{\partial y_m}e_m + \frac{\partial x_m}{\partial y_i}e_i = g_i g_m$
41. $\overline{T}$ is a transformation that transforms the vector of A_i to KA_i, where K is a scalar. Therefore $\overline{T}$ is:
   (1) A scalar
   (2) A two-order matrix
   (3) A vector
   (4) not a matrix
42. If the relation between deformed volume and initial volume is defined by $dv = JdV$ then what is the relation between the deformed surface vector and the initial surface vector?$\underline{n}$and$\underline{N}$ are the normal vectors of the deformed and initial surfaces. F is deformation gradient tensor
   (1) $\underline{N}\,\mathrm{dA} = \mathrm{JF}^T\underline{n}da$
   (2) $d\underline{A} = JF^{-T}d\underline{a}$
   (3) $Jd\underline{a} = F^T d\underline{A}$
   (4) $Jd\underline{A} = F^T d\underline{a}$
43. What is the relationship between the second invariant of deviator stress and octahedral shear stress?
   (1) $\mathrm{I}_{\Sigma D} = -\frac{3}{2}\tau_{oct}$
   (2) $\mathrm{I}_{\Sigma D} = -\frac{3}{2}\tau_{oct}^2$
   (3) $\mathrm{I}_{\Sigma D} = -\frac{1}{2}\tau_{oct}^2$
   (4) $\mathrm{I}_{\Sigma D} = -\frac{1}{2}\tau_{oct}$
44. The Cauchy stress tensor of a body is shown for equilibrium position and there is no volume force, which item is true?

$$\sigma = \begin{bmatrix} -Ax_2 & Bx_1 & \circ \\ Bx_1 & Cx_1 + Dx_2 & \circ \\ \circ & \circ & Dx_1 - Bx_2 \end{bmatrix}$$

   (1) $D = -C$
   (2) $D = C$
   (3) $B = D$
   (4) $B = -D$
45. Find the value of $\varepsilon_{1jk}A_2B_{kj}$
   (1) $A_2B_{32} - A_2B_{23}$
   (2) $A_2B_{23} - A_2B_{32}$
   (3) $A_2B_{13} - A_1B_{31}$
   (4) 0

46. What are free indices in the equation?$a_{ij}b^{jk} - r_{il}s^{lmn}t_nV_m^k = 0$
    (1) k, i
    (2) k, j
    (3) l, i
    (4) n, i
47. Find the deformed surface vendor for the following deformation if the initial surface vector is e_1.

$$x_1 = X_1 - 2X_2x_2 = X_2x_3 = X_3$$

    (1) $e_1 - 2e_2$
    (2) $e_1 - 2e_2$
    (3) $e_2 - 2e_1$
    (4) $e_2 - 2e_1$
48. Which item is true for following stress tensor?

$$T_{ij} = \begin{bmatrix} 3 & -1 & 0 \\ -1 & 3 & 0 \\ 0 & 0 & 1 \end{bmatrix}$$

    (1) Maximum normal stress is 3
    (2) Maximum shear stress is 1.5
    (3) 3)The hydrostatic stress is 3
    (4) Maximum normal stress is twice the hydrostatic stress
49. Find the relationship between a, b, c, and d for an incompressible and non-rotational flow if the Eulerian velocity field is as below:, $v_2 = ax_1 + dx_2$, $v_3 = 0v_1 = ax_1 + bx_2$
    (1) $a = d, b = -c$
    (2) $a = -d, b = c$
    (3) $a = -d, b = -c$
    (4) $a = d, b = c$
50. For below Eulerian velocity field given below, what is the second component of acceleration a_2?

$$v_1 = 2x_1t - 3x_2v_2 = 3x_2 + 2tx_1v_3 = 0$$

    (1) 0
    (2) $2x_1$
    (3) $4x_1t^3 - 6x_2t + 6tx_1 + 9x_2 + 2x_1$
    (4) $4x_1t^2 - 6x_2t + 6tx_1 + 9x_2$
51. The deformation field for a cylinder is defined below where μ, β, and ν are constant. What is the relation between these constants if the deformation is incompressible?

$$x_1 = \mu[X_1 cos(\beta X_3) + X_2 sin(\beta X_3)]$$

$$x_2 = \mu[-X_1 sin(\beta X_3) + X_2 cos(\beta X_3)]$$

$$x_3 = vX_3$$

(1) $v\mu^2 = 1$
(2) $v\mu = \beta$
(3) $v\mu^3 = \beta$
(4) $v\beta^2 = \mu$

52. For below given stress tensor at point P which is on a plane, find the value of σ_{11} on the condition that the stress vector applied on the plane at P is zero.

$$[\sigma_{ij}] = \begin{bmatrix} \sigma_{11} & 2 & 1 \\ 2 & 0 & 2 \\ 1 & 2 & 0 \end{bmatrix}$$

(1) 1
(2) 2
(3) 3
(4) 4

53. Two forms of Hook law are defined as below. Which item is true?

$$e_{ij} = \frac{1+v}{E}\sigma_{ij} - \frac{v}{E}\sigma_{kk}\delta_{ij}$$

$$\sigma_{ij} = \lambda e_{kk}\delta_{ij} + 2\mu e_{ij}$$

(1) $\frac{Ev}{(1-v)(1+2v)}$
(2) $\frac{Ev}{(1+v)(1-2v)}$
(3) $\frac{Ev}{(1+v)(1+2v)}$
(4) $\frac{Ev}{(1-v)(1-2v)}$

54. For the below stress tensor at point P, what is the ratio of shear stress to normal stress in a plane with normal vector $i_1 + i_2 + i_3$

$$T = \begin{bmatrix} 1 & 1 & 1 \\ 1 & 1 & 1 \\ 1 & 1 & 1 \end{bmatrix}$$

(1) 2
(2) 1

(3) 0
(4) 3

55. τ_{ij} is an axisymmetric stress tensor and is e_i the unit vector of Cartesian coordinates, find the value of $(e_i \times e_j)\tau_{ij}$
(1) τ_{ij}
(2) $(e_i.e_j)\tau_{ij}$
(3) 0
(4) 1

56. For the deformation gradient tensor, which is given below, what is the required condition for a plane deformation?

$$F = \begin{bmatrix} \text{a} & b & 0 \\ c & d & 0 \\ 0 & 0 & e \end{bmatrix}$$

(1) $e = 0$
(2) $det(F) = 1$
(3) $ad - bc = 1$
(4) $e = 1$

57. Stress tensor is applied on a point which is on the plane $6x_1 + 2x_2 + 3x_3 = 12$. find the stress vector acting on the plane at the point.

$$[T] = \begin{bmatrix} 14 & 7 & -7 \\ 7 & 2 & 0 \\ -7 & 0 & 35 \end{bmatrix}$$

(1) $\vec{t} = 95\vec{e_1} + 63\vec{e_3}$
(2) $\vec{t} = 6\vec{e_1} + 12\vec{e_2} + 9\vec{e_3}$
(3) $\vec{t} = 95\vec{e_1} + 84\vec{e_2} + 63\vec{e_3}$
(4) $\vec{t} = 11\vec{e_1} + 12\vec{e_2} + 9\vec{e_3}$

58. Find the value of x in following stress tensor if the principal stresses are given as below:

$$[\delta_{\text{ij}}] = \begin{bmatrix} 7 & \text{x} & 0 \\ \text{x} & 6 & -2 \\ 0 & -2 & 5 \end{bmatrix} Mpa \quad \begin{matrix} \delta_1 = 9Mpa \\ \delta_2 = 6Mpa \\ \delta_3 = 3Mpa \end{matrix}$$

(1) 1
(2) 2
(3) 3
(4) 4

59. If a_{ij} is a constant value, find $\left(a_{ij}x_ix_j\right)_{,k}$
   (1) $a_{ik}x_i + a_{jk}x_j$
   (2) $a_{ij}x_i + a_{jk}x_j$
   (3) $a_{jk}x_i + a_{ik}x_j$
   (4) $a_{ki}x_j + a_{ij}x_j$
60. The motion of a body is defined by following equation in Lagrangian coordinates, if the particle is at the position of (9,6,1) in t = 2 s, find its initial position.

$$x = \left(X_1 + t^2X_2\right)\overrightarrow{e_1} + \left(X_2 + t^2X_1\right)\overrightarrow{e_2} + (X_3)\overrightarrow{e_3}$$

   (1) (1,5,1)
   (2) (2,2,1)
   (3) (5,1,1)
   (4) (1,2,1)
61. For the deformation given below, find the extension of the line with direction vector of $e_1 + e_2$.

$$x_1 = X_1 + 2X_2,\ x_2 = X_2,\ x_3 = X_3.$$

   (1) $\frac{\sqrt{5}}{2}$
   (2) $\sqrt{5}$
   (3) $\sqrt{10}$
   (4) $2\sqrt{5}$
62. What is the value of k for an incompressible fluid based on following Eulerian velocity field.

$$V_1 = K(x_2 - 2)^2$$

$$V_2 = -x_1x_2$$

$$V_3 = Kx_1x_3$$

   (1) 1/2
   (2) 1
   (3) 3/2
   (4) 2

63. If the stress invariants of both stress tensors are similar, find the values of α and β.

$$\begin{bmatrix} 5 & a & -a \\ a & 0 & b \\ -a & b & 0 \end{bmatrix}, \begin{bmatrix} \alpha & 0 & c \\ 0 & 2 & 0 \\ c & 0 & \beta \end{bmatrix}$$

(1) −4, 7
(2) −4, −7
(3) 4, −7
(4) 4, 7

References

1. Mechanics of Continuous Media by M. Rahimian, M. Eskandari-Ghadi, Theran University Publication Institute, 2008, ISBN978-964-03-4014-1
2. Continuum Mechanics by A.J.M Spencer, Dover Publications, Longman, Mineola, New York, 1980, ISBN -9780486139470.
3. Notes on continuum mechanics (Lecture Notes on Numerical Methods in Engineering and Sciences Book 4) by Eduardo W.V. Chavez, 2013 Edition, Kindle Edition, Springer, ISBN-13: 978–9400759855.
4. Continuum Mechanics for Engineers by G. Thomas Mase, Ronald E. Smelser, George E. Mase, 3rd Edition, CRC Press, 2009, ISBN 9781420085389 - CAT# 85387.
5. Continuum Mechanics, by M.H. Hamedi, 2012, Andisheh Nasir Publications, ISBN: 9786009212927.
6. Fotis Sotiropoulos, Iman Borazjani, A review of state-of-the-art numerical methods for simulating flow through mechanical heart valves, Med. Biol. Eng. Comput. 47 (2009) 245–256.
7. Syed Samar Abbas, Mohammad Shakir Nasif, Rafat Al-Waked, Mior Azman Meor Said, Numerical investigation on the relationship of cavitation initiation in bileaflet mechanical heart valves (bmhvs) with the aortic root geometry and valve's implantation rotation angle, J.e Braz. Soc. Mech. Sci. Eng. 42 (2020) 1–17.
8. Nurul Aisyah Hasni, Aisyah Ahmad Shafi, Muhammad Ikman Ishak, The effect of bileaflet mechanical heart valve designs on biomechanical behaviours–a finite element analysis, Jurnal Teknologi 86 (1) (2024) 43–52
9. Ashok K Singhal, Mahesh M Athavale, Huiying Li, Yu Jiang, Mathematical basis and validation of the full cavitation model, J. Fluids Eng. 124 (3) (2002) 617–624.
10. Chee C. Y., Lee H. P., Lu C. Using 3D fluid–structure interaction model to analyze the biomechanical properties of erythrocyte. Physics Letters A,2008; 372 : 1357–1362.
11. Evans EA, Hochmuth RM. Membrane viscoelasticity. Biophys J, 1976 ;16(1):1–11.
12. Dao M., Lim C. T., Suresha S. Mechanics of the human red blood cell deformed by optical tweezers. Journal of the Mechanics and Physics of Solids. 2003; 51: 2259 – 2280.
13. Mills J. P., Qie L., Dao M., Lim, et al. Nonlinear elastic and viscoelastic deformation of the human red blood cell with optical tweezers. Mech Chem Biosyst. 2004;1(3): 169–180
14. Tomaiuolo G, Preziosi V, Simeone M, et all. A methodology to study the deformability of red blood cells flowing in microcapillaries in vitro. Ann Ist Super Sanità ,2007; 43(2): 186–192.
15. Miller G., Marquardt P., and Gleiter, H. Size-Induced Metal-Insulator Transition in Metals and Semiconductors . J. Cryst. Growth, 1992; 86(1-4): 66-71
16. McGarry J. G. and Prendergast, P. J. A three-dimensional finite element model of an adherent eukaryotic cell, European Cells and Materials, 2004; 7: 27–34).

K. Hassani and F. Azimifar, *Introduction to Continuum Mechanics*, Synthesis Lectures on Mechanical Engineering, https://doi.org/10.1007/978-3-032-05076-2

17. Tsubota K, Wada S, and Liu H. Computer simulation of tank-treading and tumbling motions of red blood cells under the influence of the natural state of an elastic cellular membrane. J. C. F. Pereira and A. Sequeira (Eds) Lisbon, Portugal, 14–17 June 2010.
18. Korin N, Bransky A,and Dinnar U.Theoretical model and experimental study of red blood cell (RBC) deformation in microchannels. Journal of Biomechanics,2007; 40 : 2088–2095.
19. Evans E.A., and Fung, Y.C. Improved measurements of the erythrocyte geometry. Microvasc Res., 1972; 4:335–347.
20. Murman S. M., Aftosmis, M. J., and Berger, M. J. Simulations of 6-DOF Motion with a Cartesian Method, 41st AIAA Aerospace Sciences Meeting Jan. 2003.
21. Salman, H.E.; Yalcin, H.C. Computational Modeling of Blood Flow Hemodynamics for Biomechanical Investigation of Cardiac Development and Disease. J. Cardiovasc. Dev. Dis. **2021**, 8, 14. https://doi.org/10.3390/jcdd8020014
22. Bazilevs, Y., Calo, V.M., Hughes, T.J.R. et al. Isogeometric fluid-structure interaction: theory, algorithms, and computations. Comput Mech **43**, 3–37 (2008). https://doi.org/10.1007/s00466-008-0315-x
23. Tezduyar, T.E., Sathe, S., Schwaab, M. and Conklin, B.S. (2008), Arterial fluid mechanics modeling with the stabilized space–time fluid–structure interaction technique. Int. J. Numer. Meth. Fluids, 57: 601-629. https://doi.org/10.1002/fld.1633

MIX
Papier aus verantwortungsvollen Quellen
Paper from responsible sources
FSC® C105338

If you have any concerns about our products,
you can contact us on
ProductSafety@springernature.com

In case Publisher is established outside the EU,
the EU authorized representative is:
Springer Nature Customer Service Center GmbH
Europaplatz 3, 69115 Heidelberg, Germany

Printed by Libri Plureos GmbH
in Hamburg, Germany